Advances in Solid Oxide Fuel Cells V

Advances in Solid Oxide Fuel Cells V

A Collection of Papers Presented at the
33rd International Conference on
Advanced Ceramics and Composites
January 18–23, 2009
Daytona Beach, Florida

Edited by
Narottam P. Bansal
Prabhakar Singh

Volume Editors
Dileep Singh
Jonathan Salem

The
American
Ceramic
Society

WILEY

A John Wiley & Sons, Inc., Publication

Published by John Wiley & Sons, Inc., Hoboken, New Jersey.
Published simultaneously in Canada.

For general information on our other products and services or for technical support, please contact our
Customer Care Department within the United States at (800) 762-2974, outside the United States at
(317) 572-3993 or fax (317) 572-4002.

Wiley also publishes its books in a variety of electronic formats. Some content that appears in print may
not be available in electronic format. For information about Wiley products, visit our web site at
www.wiley.com.

Library of Congress Cataloging-in-Publication Data is available.

ISBN 978-0-470-45754-2

10 9 8 7 6 5 4 3 2 1

Contents

ELECTRODES

OXIDE/PROTON CONDUCTORS

SEALS

MECHANICAL BEHAVIOR

MATERIALS SYNTHESIS

FUEL REFORMING

Preface

The Sixth International Symposium on Solid Oxide Fuel Cells (SOFC): Materials, Science, and Technology was held during the 33rd International Conference and Exposition on Advanced Ceramics and Composites in Daytona Beach, FL, January 18 to 23, 2009. This symposium provided an international forum for scientists, engineers, and technologists to discuss and exchange state-of-the-art ideas, information, and technology on various aspects of solid oxide fuel cells. A total of 112 papers were presented in the form of oral and poster presentations, including ten invited lectures, indicating strong interest in the scientifically and technologically important field of solid oxide fuel cells. Authors from 17 countries (Brazil, Canada, China, Denmark, France, Georgia, Germany, India, Italy, Japan, Romania, Russia, Singapore, South Korea, Taiwan, United Kingdom, and U.S.A.) participated. The speakers represented universities, industries, and government research laboratories.

These proceedings contain contributions on various aspects of solid oxide fuel cells that were discussed at the symposium. Twenty four papers describing the current status of solid oxide fuel cells technology and the latest developments in the areas of fabrication, characterization, testing, performance, electrodes, electrolytes, seals, cell and stack development, proton conductors, fuel reforming, mechanical behavior, powder synthesis, etc. are included in this volume. Each manuscript was peer-reviewed using The American Ceramic Society review process.

The editors wish to extend their gratitude and appreciation to all the authors for their contributions and cooperation, to all the participants and session chairs for their time and efforts, and to all the reviewers for their useful comments and suggestions. Financial support from The American Ceramic Society is gratefully acknowledged. Thanks are due to the staff of the meetings and publications departments of The American Ceramic Society for their invaluable assistance. Advice, help and cooperation of the members of the symposium's international organizing committee (Tatsumi Ishihara, Tatsuya Kawada, Nguyen Minh, Mogens Mogensen, Nigel Sammes, Robert Steinberger-Wilkens, Jeffry Stevenson, and Eric Wachsman) at various stages were instrumental in making this symposium a great success.

It is our earnest hope that this volume will serve as a valuable reference for the engineers, scientists, researchers and others interested in the materials, science and technology of solid oxide fuel cells.

NAROTTAM P. BANSAL
NASA Glenn Research Center

PRABHAKAR SINGH
University of Connecticut

Introduction

The theme of international participation continued at the 33rd International Conference on Advanced Ceramics and Composites (ICACC), with over 1000 attendees from 39 countries. China has become a more significant participant in the program with 15 contributed papers and the presentation of the 2009 Engineering Ceramic Division's Bridge Building Award lecture. The 2009 meeting was organized in conjunction with the Electronics Division and the Nuclear and Environmental Technology Division.

Energy related themes were a mainstay, with symposia on nuclear energy, solid oxide fuel cells, materials for thermal-to-electric energy conversion, and thermal barrier coatings participating along with the traditional themes of armor, mechanical properties, and porous ceramics. Newer themes included nano-structured materials, advanced manufacturing, and bioceramics. Once again the conference included topics ranging from ceramic nanomaterials to structural reliability of ceramic components, demonstrating the linkage between materials science developments at the atomic level and macro-level structural applications. Symposium on Nanostructured Materials and Nanocomposites was held in honor of Prof. Koichi Niihara and recognized the significant contributions made by him. The conference was organized into the following symposia and focused sessions:

Symposium 1	Mechanical Behavior and Performance of Ceramics and Composites
Symposium 2	Advanced Ceramic Coatings for Structural, Environmental, and Functional Applications
Symposium 3	6th International Symposium on Solid Oxide Fuel Cells (SOFC): Materials, Science, and Technology
Symposium 4	Armor Ceramics
Symposium 5	Next Generation Bioceramics
Symposium 6	Key Materials and Technologies for Efficient Direct Thermal-to-Electrical Conversion
Symposium 7	3rd International Symposium on Nanostructured Materials and Nanocomposites: In Honor of Professor Koichi Niihara
Symposium 8	3rd International symposium on Advanced Processing & Manufacturing Technologies (APMT) for Structural & Multifunctional Materials and Systems

Symposium 9	Porous Ceramics: Novel Developments and Applications
Symposium 10	International Symposium on Silicon Carbide and Carbon-Based Materials for Fusion and Advanced Nuclear Energy Applications
Symposium 11	Symposium on Advanced Dielectrics, Piezoelectric, Ferroelectric, and Multiferroic Materials
Focused Session 1	Geopolymers and other Inorganic Polymers
Focused Session 2	Materials for Solid State Lighting
Focused Session 3	Advanced Sensor Technology for High-Temperature Applications
Focused Session 4	Processing and Properties of Nuclear Fuels and Wastes

The conference proceedings compiles peer reviewed papers from the above symposia and focused sessions into 9 issues of the 2009 Ceramic Engineering & Science Proceedings (CESP); Volume 30, Issues 2-10, 2009 as outlined below:

- Mechanical Properties and Performance of Engineering Ceramics and Composites IV, CESP Volume 30, Issue 2 (includes papers from Symp. 1 and FS 1)
- Advanced Ceramic Coatings and Interfaces IV Volume 30, Issue 3 (includes papers from Symp. 2)
- Advances in Solid Oxide Fuel Cells V, CESP Volume 30, Issue 4 (includes papers from Symp. 3)
- Advances in Ceramic Armor V, CESP Volume 30, Issue 5 (includes papers from Symp. 4)
- Advances in Bioceramics and Porous Ceramics II, CESP Volume 30, Issue 6 (includes papers from Symp. 5 and Symp. 9)
- Nanostructured Materials and Nanotechnology III, CESP Volume 30, Issue 7 (includes papers from Symp. 7)
- Advanced Processing and Manufacturing Technologies for Structural and Multifunctional Materials III, CESP Volume 30, Issue 8 (includes papers from Symp. 8)
- Advances in Electronic Ceramics II, CESP Volume 30, Issue 9 (includes papers from Symp. 11, Symp. 6, FS 2 and FS 3)
- Ceramics in Nuclear Applications, CESP Volume 30, Issue 10 (includes papers from Symp. 10 and FS 4)

The organization of the Daytona Beach meeting and the publication of these proceedings were possible thanks to the professional staff of The American Ceramic Society (ACerS) and the tireless dedication of the many members of the ACerS Engineering Ceramics, Nuclear & Environmental Technology and Electronics Divisions. We would especially like to express our sincere thanks to the symposia organizers, session chairs, presenters and conference attendees, for their efforts and enthusiastic participation in the vibrant and cutting-edge conference.

DILEEP SINGH and JONATHAN SALEM
Volume Editors

Cell and Stack
Development/Performance

DEVELOPMENT OF NOVEL PLANAR NANO-STRUCTURED SOFCs

Tim Van Gestel, Doris Sebold, Wilhelm A. Meulenberg, Hans-Peter Buchkremer
Forschungszentrum Jülich GmbH, Institute of Energy Research, IEF-1: Materials Synthesis
and Processing, Leo-Brandt-Strasse, D-52425 Jülich, Germany

ABSTRACT
This paper reports a study on the preparation of thin nano-structured 8YSZ membrane films by nano-dispersion and sol-gel coating methods. For the deposition process, four different coating liquids with varying particle size, covering the range from 85 nm to 6 nm, were prepared. In the first part, it is demonstrated that nano-dispersions with a particle size of e.g. 85 or 65 nm can be used for the formation of dense YSZ membranes with a thickness of 1 – 2 μm, but the layers can only be sintered to full density at 1400°C. In the second part, ultra-thin YSZ membranes are prepared by sol-gel coating (particle size 35 and 6 nm). These layers show a thickness < 500 nm and a very tight mesoporous or microporous structure in the calcined state (pore size < 5 nm), which leads to sintered membrane layers in the desired temperature range (~ 1200°C). In the final part, it is described by means of He leak tests that on the applied large-scale state-of-the-art substrate with a high surface roughness, a firing temperature of 1300°C is however required to sinter the membranes to full density.

INTRODUCTION

At Forschungszentrum Jülich (FZJ), an advanced planar anode supported SOFC has been developed, characterized by an average power output of 1.4 W/cm^2 at 750°C and 0.7 V. The typical multilayer structure of the FZJ cell comprises: (1) a porous 8YSZ/NiO anode substrate with an area of 20 x 20 cm^2; (2) a porous 8YSZ/NiO anode functional layer; (3) a dense 8YSZ electrolyte layer with a thickness in the range 5 – 10 μm; (4) a porous LSM or LSCF cathode. In the standard manufacturing procedures, the substrate is made by warm pressing or tape-casting and, subsequently, the respective layers are deposited from a suspension by means of vacuum slip-casting, screen-printing or wet-powder spraying.

Currently, one of the main objectives in our research is to produce novel electrolyte membrane layers, which can be sintered to full density at a lower temperature than currently applied in SOFC manufacturing (> 1400°C). This will reduce significantly the production cost of the cell and permit the use of steel substrates to build up the cell.

In this paper, nano-dispersion and sol-gel coating procedures are described and their potential for SOFC manufacturing is discussed. In our related research group 'separation membranes', such coating procedures are already used for several years as a method of producing thin mesoporous (pore size > 2 nm) and microporous (pore size < 2 nm) membrane films of several materials including electrolyte materials (e.g. 3Y$_2$O$_3$-ZrO$_2$, 8Y$_2$O$_3$-ZrO$_2$, 10Gd$_2$O$_3$-CeO$_2$) [1]. The pore size of such membrane films can be varied in the range from ~ 20 nm down to even smaller than 1 nm - depending on the particle size of the coating liquid - and is considerably smaller than the pore size of a macroporous membrane films made by a conventionally used suspension deposition procedure.

From a practical point of view, coating of sols and nano-dispersions appears also as an interesting method for SOFC manufacturing, since the coating methods are the same as conventional suspension methods and exhibit the same advantages (inexpensive in terms of capital costs, simplicity of the equipment). The basic question when considering these new thin-film coating methods includes, however, how to deal with a regular anode substrate showing an inferior surface quality when compared with substrates usually used for thin film deposition. As shown in Figure 1, the thickness of the 8YSZ electrolyte layer, which is normally deposited on our anode layer by a powder procedure, measures ~ 10 μm after sintering.

Fig. 1. SEM micrograph of the standard SOFC prepared at FZJ with 8YSZ/NiO anode substrate, 8YSZ/NiO anode layer, 8YSZ electrolyte, 8YSZ/LSM cathode layer and LSM cathode (bar = 10 μm).

EXPERIMENTAL

1. Anode Substrate

The standard anode produced at our institute (IEF-1), which consists of a warm-pressed 8YSZ/NiO plate and a relatively thin vacuum slip-casted macroporous 8YSZ/NiO layer, is used as substrate in all coating experiments. From Figure 1, it appears that the thickness of the anode layer measures ~ 10 μm; the thickness of the supporting plates was 1 mm or 1.5 mm in all experiments.

In Figure 2a, the pore size distribution of the substrate is shown. From this graph, an average pore size of ~ 0.7 – 0.8 μm (large peak) and ~ 200 – 300 nm (small peaks) is evident for the substrate plate and for the anode layer, respectively.

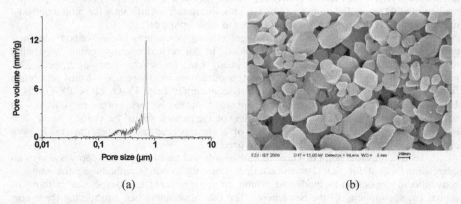

(a)   (b)

Fig. 2a. Pore size distribution of test substrate consisting of a 8YSZ/NiO anode plate and a 8YSZ/NiO anode layer. Figure 2b. Surface micrograph of the anode layer.

Figure 2b shows a detail surface micrograph of the anode layer. This micrograph confirms a typical macroporous structure with a particle size in the range 200 – 400 nm and an average pore size in the range 200 – 300 nm. Further, it is also confirmed that the anode layer shows a wide pore size distribution.

2. Preparation of the coating liquids

As shown in previous coating experiments, deposition of continuous membrane layers on this kind of anode requires a coating liquid containing relatively large particles [2]. In a series of coating experiments, dense 8YSZ electrolyte layers with a thickness in the range 1 - 4 μm could be deposited, when the coating liquid was a dispersion with a particle size of ~ 200 nm and PVA was added as coating additive (coating methods included dip-coating, vacuum-casting and spraying). The main drawback of our proposed coating procedure was however the requirement of a high sintering temperature of 1400°C for these 8YSZ membranes, similar to the sintering temperature of conventional electrolyte layers. In this paper, a new series of coating experiments with four different coating liquids - including two nano-dispersions and two sols - is described, in order to check the effect of a decreasing particle size on the sintering behaviour of the electrolyte membrane layer.

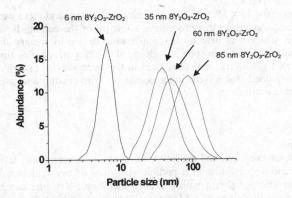

Fig. 3. Particle size distribution of nano-dispersions and sols descirbed in this work

A first nano-dispersion was made starting from a commercially available 8YSZ nano-powder (Evonik Degussa) and an aqueous nitric acid solution. Characterization of the particle size was done by dynamic laser beam scattering (Horiba LB-550) in a similar way as in our previous work (1) and here an average particle size of ~ 85 nm was measured. A coating liquid was prepared from this dispersion by adding polyvinyl alcohol (PVA) as coating and drying controlling additive. Subsequently, a second nano-dispersion with a particle size of ~ 65 nm was prepared, using a 8YSZ nano-powder supplied by Sigma-Aldrich.

The third coating liquid was a 'so-called' colloidal sol. The sol was prepared starting from a metal organic precursor ($Zr(n-OC_3H_7)_4$, Aldrich), by means of hydrolysis and condensation of this precursor in isopropanol-water-HNO_3 solutions at 98°C. Yttria-doping (8 mol% yttria) was done by adding the proper amount of $Y(NO_3)_3.6H_2O$ to the zirconia sol. In this procedure, the size of the particles in the sol was controlled by the time of the aging process and the best results were obtained for particles with an average size of ~ 30 – 40 nm.

The fourth coating liquid in this study was a 'so-called' polymeric sol, which contains particles with a size in the nanometer range. This sol was produced by controlled hydrolysis of $Zr(n-OC_3H_7)_4$, in the presence of diethanol amine (DEA) as a precursor modifier/polymerization inhibitor and an yttrium precursor $(Y(i-OC_3H_7)_3)$ as a doping compound. The essential feature of the preparation route is the addition of DEA, which leads to a reproducible formation of nano-particles with a size of 5 - 10 nm in the synthesis process and also acts as a coating and drying controlling additive [3]. In Figure 3, an overview of the size distributions measured by dynamic laser scattering of the prepared sols and nano-dispersions is shown.

3. Membrane coating

Dip-coating experiments were performed using an automatic dip-coating device, equipped with a holder for 4 x 4 cm^2 substrates. The anode substrates were cut from calcined anode plates obtained from our standard production process (calcination temperature 1000°C, dimension 25 x 25 cm^2). In the dip-coating process, sol particles were deposited as a membrane film by contacting the upper-side (anode layer side) of the substrate with the coating liquid, while a small under-pressure was applied at the back-side. The obtained supported gel-layers were fired in air at 500°C for 2 h and then the coating – calcination cycle was repeated once, unless stated otherwise. A final sintering treatment was carried out at a temperature of 1400°C, 1300°C or 1200°C.

For He leak test measurements, 75 x 75 mm^2 anode substrates were cut and these were coated by means of a spin-coating device. In this procedure, a few ml of the coating liquid was dropped onto the substrate, which was held to the spin-coating device by means of a vacuum holder. After 30s, the substrate was then spun at high speed (1200 rpm, spinning time 1 minute). The obtained supported gel-layers were then fired in the same way as described above. Further, it should be mentioned that all coating procedures – dip-coating and spin-coating – were carried out in a clean-room.

RESULTS AND DISCUSSION
1. 85 nm 8YSZ Nano-dispersion

Figures 4a and 4b show an overview and a detail micrograph of a membrane layer obtained by coating the first nano-dispersion with a particle size of ~ 85 nm (calcination temperature 500°C). From these micrographs, a typical graded membrane structure can be observed comprising subsequently the calcined 8YSZ layer, the macroporous anode layer and the supporting anode plate. The coated 8YSZ layer is in this back-scattering type of SEM micrograph visible as a brighter film, due to its much smaller pore size. Further, it appears that a separate and continuous layer was obtained, which uniformly covers the anode layer.

By looking at the detail micrograph 4b, it appears clearly that infiltration of 8YSZ particles into the macropores of the anode layer could be prevented, by applying a plastic compound (PVA) as an additive in the coating liquid. In this micrograph, a separation line between the successive membrane layers is also visible and it appears that the thickness of a single layer obtained by one coating – calcination step measures ~ 2 – 3 μm.

The evolution of the structure of the calcined 8YSZ membrane layer at higher temperatures is shown in Figures 4c – 4f. According to the first cross-section micrograph in 4c, a dense membrane layer was obtained at a reduced firing temperature of 1300°C. However, after examanition of the surface micrographs, it appeared that densification had only proceeded to some extent (Figure 4d).

In the micrographs taken after firing at 1400 °C, a fully sintered layer was observed and much larger sintering grains are visible (Figure 4f). From all these micrographs, it was thus concluded that the coating procedure was effective - very thin dense electrolyte films with a thickness of ~ 2 μm were obtained - but our attempt to make a dense electrolyte membrane layer at a temperature < 1400°C was not succesfull.

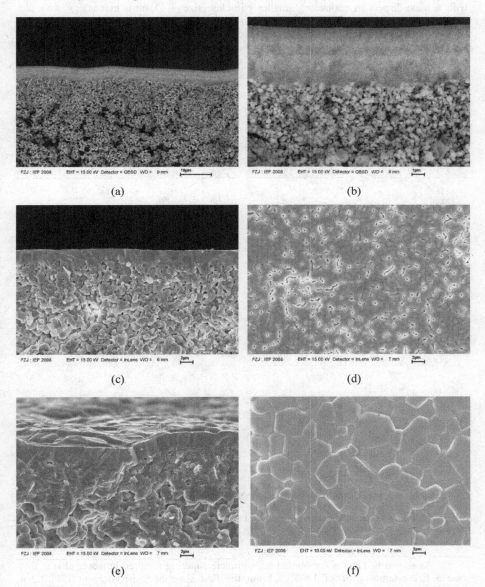

(a)

(b)

(c)

(d)

(e)

(f)

Fig. 4a and 4b. Micrographs of a calcined mesoporous $8Y_2O_3$-ZrO_2 membrane layer, obtained by dip-coating nano-dispersion 1 on the standard anode substrate. Figure 4c and 4d. Micrographs of the membrane after firing at 1300°C. Fig. 4e and 4f.. Micrographs of the membrane after firing at 1400°C.
(Layers made by 2 x dip-coating and calcination at 500°C; (a) bar = 10 μm, (b) bar = 1 μm, (c,d,e,f,) bar = 2 μm)

2. 65 nm 8YSZ Nano-dispersion

Figure 5 shows micrographs of a calcined 8YSZ membrane layer which was prepared with a nano-dispersion containing smaller particles (size ~ 65 nm). In analogy with the previous coating experiment, a continuous membrane layer was formed and infiltration of the smaller particles into the pores of the macroporous anode layer did also not occur.

The main differences with the previous coating experiment included a decreased overall membrane thickness in the calcined state of ~ 2 µm and - as shown in surface micrographs 5c and 5d - a calcined layer with a significantly finer mesoporous structure was obtained.

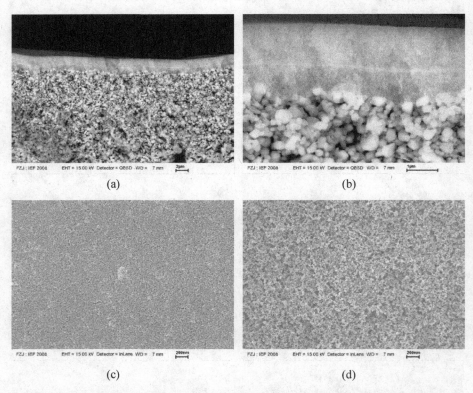

(a)

(b)

(c)

(d)

Fig. 5a and 5b. Micrographs of the cross-section of a 8YSZ membrane, obtained by dip-coating nano-dispersion 2 on the standard anode substrate. Fig. 5c and 5d. Comparison of the surface of 8YSZ membranes, obtained by dip-coating nano-dispersion 2 (5c) and nano-dispersion 1 (5d).
(Layers made by 2 x dip-coating and calcination at 500°C; (a) bar = 2 µm, (b) bar = 1, (c,d) bar = 200 nm)

As shown in Figures 6a, 6b and 6c, complete sintering required however also in this case a firing temperature of 1400°C. From the first sintering experiment at 1200°C, it appeared that densification had already proceeded to a larger extent when compared with the previous coating experiment, but the layer still contained a significant amount of pores. Further, as shown in Figure 6a (d), the membrane layer still contained a number of locally untight areas, especially at places where the surface of the substrate shows strong curvatures.

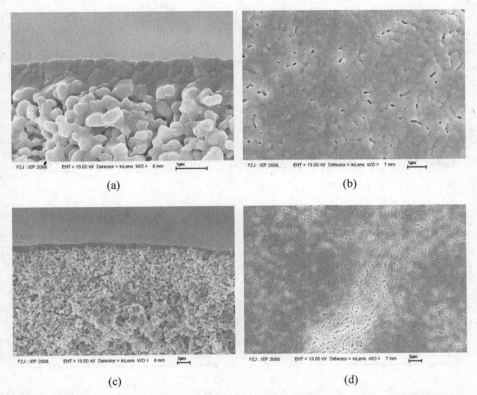

Fig. 6a (a) – (d). Micrographs of the cross-section and surface of the membrane shown in Figure 5 after firing at 1200°C.
(Layers made by 2 x dip-coating, calcination at 500°C; (a,b) bar = 1 μm, (c,d) bar = 2 μm)

After firing at 1300°C, larger grains were formed and a sintered layer with a thickness of ~ 1 μm was observed, which looked fully sintered in the detail SEM surface micrographs (Figures 6b (a)-(d)). However, after checking overview surface micrographs taken at a lower magnification (6b (e)), we still could observe a number of untight areas. In Figure 6b (f), an example of such an area is shown.

In the micrographs taken after firing at 1400°C, larger sintering grains became visible and the layer appeared completely dense, also in overview SEM pictures. From the cross-section picture in 6c (a), a final thickness of ~ 1 μm can be estimated, while the second picture of the layer at a smaller magnification in 6c (c) shows a somewhat larger thickness of ~ 1.5 μm. Anyway, it was again concluded that the coating procedure was effective - very thin dense electrolyte films with a thickness of 1 – 1.5 μm were obtained - but also this attempt to make a dense electrolyte membrane layer at a temperature < 1400°C was not succesfull. It should also be noticed that also the anode layer appears as a dense sintered layer in these SEM pictures, but He leak tests showed afterwards that the anode is only partially sintered. Further, before the finished cells are send for cell measurements, we also subject them to an additional firing in reduced atmosphere.

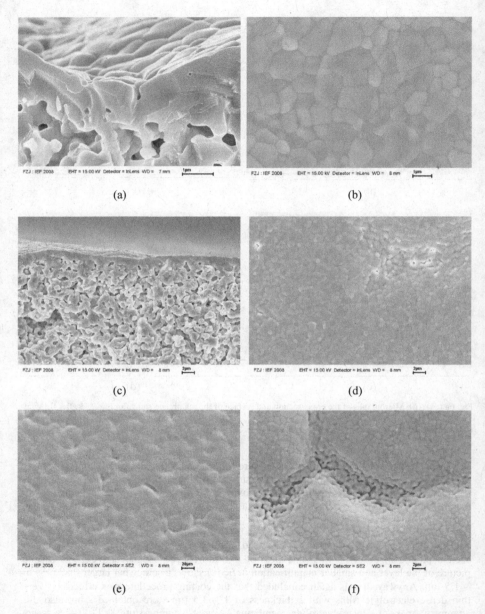

Fig. 6b (a) – (f). Micrographs of the cross-section and surface of the membrane shown in Figure 5 after firing at 1300°C
(Layers made by 2 x dip-coating, calcination at 500°C; (a,b) bar = 1 μm, (c,d,f) bar = 2 μm, (e) bar = 20 μm)

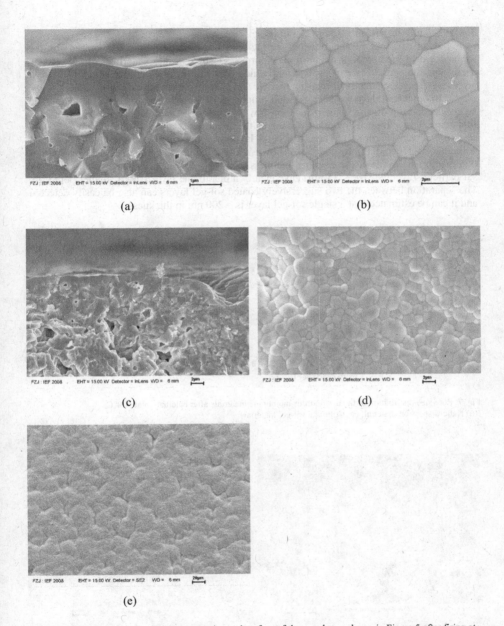

(a)

(b)

(c)

(d)

(e)

Fig. 6c (a) – (e). Micrographs of the cross-section and surface of the membrane shown in Figure 5 after firing at 1400°C.
(Layers made by 2 x dip-coating, calcination at 500°C; (a,b) bar = 1 μm, (c,d) bar = 2 μm, (e) bar = 20 μm)

3. 8YSZ Nano-dispersion + 35 nm Colloidal sol

After analysis of the previous results - a large particle size is required for the formation of a membrane layer on a very porous substrate, while sintering is improved by decreasing the particle size of the coating liquid - graded membranes, consisting of different layers with a systematically decreasing particle and pore size were developed.

In order to reduce the pore size and roughness of the substrate, a first YSZ membrane layer was deposited using nano-dispersion 2 with an average particle size of ~ 65 nm. This first layer shows a much smoother surface as the anode layer and a strongly decreased pore size of ~ 7 nm (Figure 7). Then, a second layer was deposited according to a so-called colloidal sol-gel coating procedure. As shown in the micrograph given in Figure 8a, a rather thin membrane layer with an average thickness of ~ 0.3 – 0.4 μm was obtained and the finer sol particles (average size ~ 35 nm) gave clearly a membrane layer with a smaller pore size. The separation between the two successively coated sol-gel layers can also be easily detected and it can be estimated that a single sol-gel layer is ~ 200 nm in thickness.

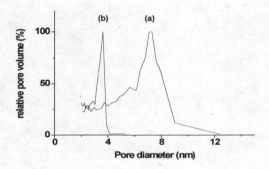

Fig. 7. Pore size distribution of the mesoporous membrane materials after calcination at 500°C. ((a) Nano-dispersion material; (b) Colloidal sol-gel material)

(a)   (b)

Fig. 8a and 8b. Cross-section and surface of a mesoporous 8YSZ membrane, obtained by dip-coating subsequently nano-dispersion 2 and a colloidal sol.
(Layers made by 2 x dip-coating and calcination at 500°C; (a) bar = 1 μm, (b) bar = 200 nm)

Pore analysis of the sol-gel membrane material with N_2-adsorption/desorption measurements indicated a mesoporous structure (type IV isotherm) with a pore size maximum of ~ 3 - 4 nm (Figure 7), while the pore size of the first layer measures ~ 7 nm. By comparing the surface of the first layer in Figure 5c and this of the second sol-gel derived layer in Figure 8b, a clear reduction in pore size and particle size after deposition of the sol-gel layer was also confirmed.

Figure 9 shows cross-section and surface micrographs of the graded membrane after further firing at 1200°C, 1300°C and 1400°C. From the first sintering experiment at 1200°C, it appeared clearly that densification had proceeded to a much larger extent when compared with the previous coating experiments and in the surface micrograph the layer looked almost completely dense. Probably, membrane coating on a substrate with a lower roughness could give SEM pictures with a completely dense layer, using this coating procedure.

As shown in Figures 9b (a) – (e), the same conclusions can be drawn for the sintering experiment at 1300°C. In the next sintering experiment at 1400°C, larger sintering grains were formed and a completely dense layer was obtained, which looked also completely tight in overview SEM pictures (Figure 9c (e)). The final thickness of the finished sintered layer measures ~ 1 – 1.5 µm .

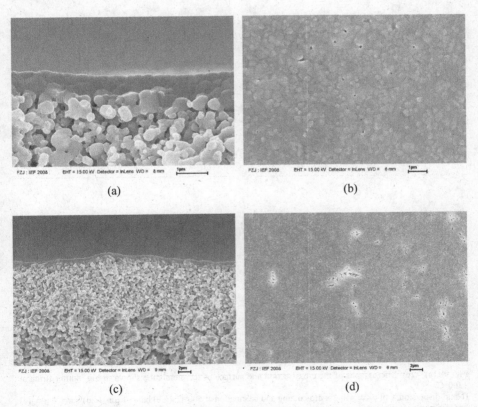

(a)

(b)

(c)

(d)

Fig. 9a (a) – (d). Micrographs of the cross-section and surface of the membrane shown in Fig. 8 after firing at 1200°C.
(Nano-dispersion, Colloidal sol: 2 x dip-coating and calcination at 500°C; (a,b) bar = 1 µm, (c,d) bar = 2 µm)

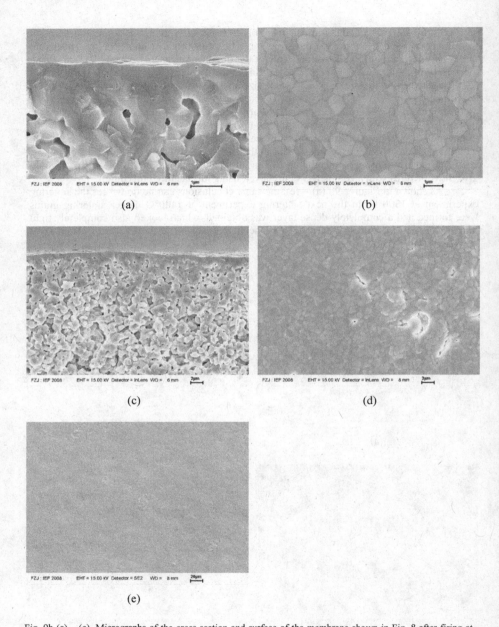

(a)   (b)

(c)   (d)

(e)

Fig. 9b (a) – (e). Micrographs of the cross-section and surface of the membrane shown in Fig. 8 after firing at 1300°C.
(Nano-dispersion, Colloidal sol: 2 x dip-coating and calcination at 500°C; (a,b) bar = 1 µm, (c,d) bar = 2 µm, (e) bar = 20 µm)

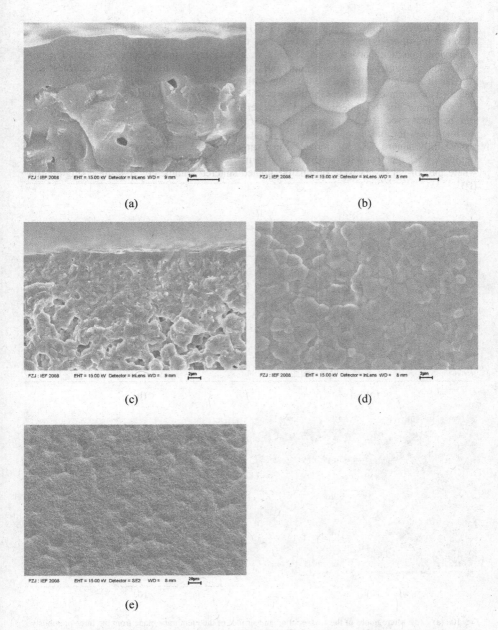

Fig. 9c (a) – (e). Micrographs of the cross-section and surface of the membrane shown in Fig. 8 after firing at 1400°C.
(Nano-dispersion, Colloidal sol: 2 x dip-coating and calcination at 500°C; (a,b) bar = 1 μm, (c,d) bar = 2 μm, (e) bar = 20 μm)

After obtaining the first results from He-leak tests, a graded combination of three different mesoporous layers was also tested, with a first layer made from the first nano-dispersion (average particle size 85 nm), a second layer made from the second nano-dispersion (average particle size 65 nm) and a third layer made from the 35 nm colloidal sol. As shown in Figures 10a – 10c, the main difference with the previous coating experiment included an increased thickness for the layers fired at 1300°C and 1400°C.

Further, for the sintering experiment performed at 1200°C, an interesting observation can be made (see for example Figure 10a (a)). Namely, three different regions can be distinguished: the macroporous anode layer, a partially sintered YSZ layer and a very thin almost fully sinterd YSZ membrane layer. Apparently, the first layer(s) made from coating liquids with larger particles sintered only partially, while the upper part of the membrane where the sol-gel layers were originally situated sintered completely at 1200°C (Figure 10a (f)).

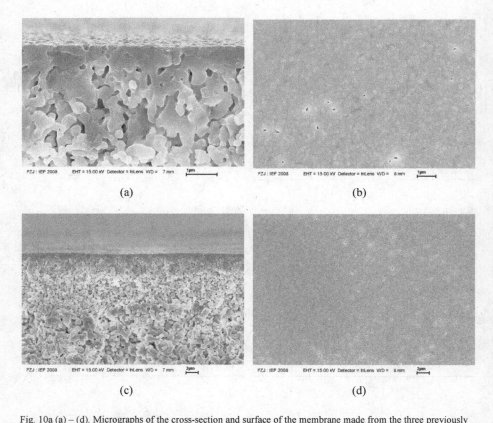

(a)

(b)

(c)

(d)

Fig. 10a (a) – (d). Micrographs of the cross-section and surface of the membrane made from the three previously discussed mesoporous membrane layers after firing at 1200°C.
(Nano-dispersion 1, Nano-dispersion 2, Colloidal sol: 1 x dip-coating and calcination at 500°C; (a,b) bar = 1 μm, (c,d) bar = 2 μm)

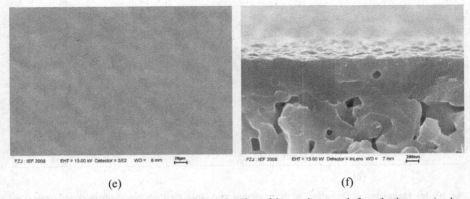

Fig. 10a (e) – (f). Micrographs of the cross-section and surface of the membrane made from the three previously discussed mesoporous layers after firing at 1200°C. (Nano-dispersion 1, Nano-dispersion 2, Colloidal sol: 1 x dip-coating and calcination at 500°C; (e) bar = 20 μm, (f) bar = 200 nm)

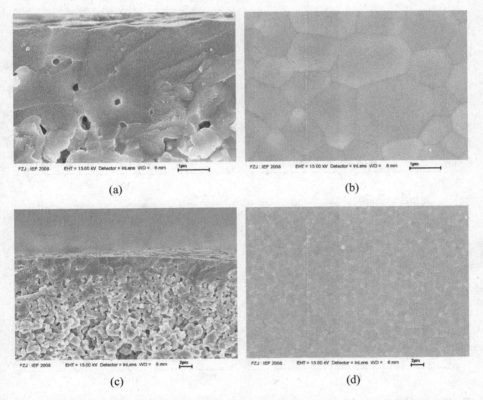

Fig. 10b (a) – (d). Micrographs of the cross-section and surface of the membrane made from the three previously discussed mesoporous membrane layers after firing at 1300°C. (Nano-dispersion 1, Nano-dispersion 2, Colloidal sol: 1 x dip-coating and calcination at 500°C; (a,b) bar = 1 μm, (c,d) bar = 2 μm)

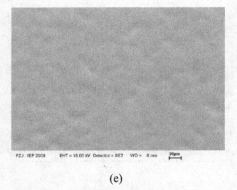

(e)

Fig. 10b (e). Micrograph of the surface of the membrane made from the three previously discussed mesoporous layers after firing at 1300°C. (Nano-dispersion 1, Nano-dispersion 2, Colloidal sol: 1 x dip-coating and calcination at 500°C; (e) bar = 20 μm)

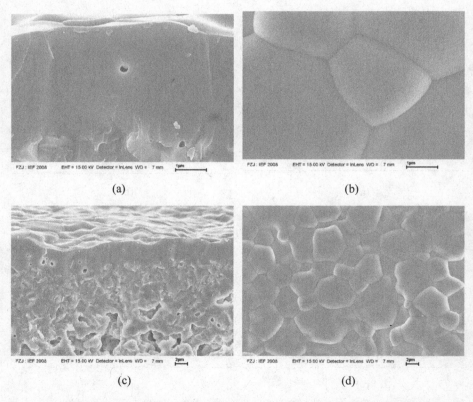

(a)   (b)

(c)   (d)

Fig. 10c (a) – (d). Micrographs of the cross-section and surface of the membrane made from the three previously discussed mesoporous membrane layers after firing at 1400°C. (Nano-dispersion 1, Nano-dispersion 2, Colloidal sol: 1 x dip-coating and calcination at 500°C; (a,b) bar = 1 μm, (c,d) bar = 2 μm)

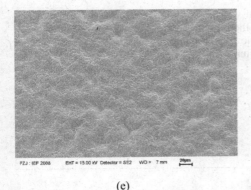

(e)

Fig. 10c (e). Micrograph of the surface of the membrane made from the three previously discussed mesoporous layers after firing at 1400°C. (Nano-dispersion 1, Nano-dispersion 2, Colloidal sol: 1 x dip-coating and calcination at 500°C; (e) bar = 20 µm)

4. 65 nm 8YSZ Nano-dispersion + 35 nm Colloidal sol + 6 nm Polymeric sol

Then, finally, a membrane conficuration with a second and even finer sol-gel layer made from nano-particles was tested. Micrographs 11a and 11b illustrate the obtained graded structure for the membrane after calcination (back-scattering mode). Successively, the following layers are present in these images: a macroporous anode layer made from a suspension with a pore size > 100 nm, a first mesoporous membrane layer made from a nano-dispersion with a pore size of ~ 7 nm, a second mesoporous membrane layer made from a colloidal sol with a pore size of ~ 3 - 4 nm and a nano-structured membrane layer made from a sol containing nano-particles. In the detail micrograph, also the separation line between the successive membrane layers is visible and it appears that the thickness of the last layer measures ~ 100 nm.

(a)   (b)

Fig. 11. Overview and detail image of the cross-section of a 8YSZ membrane, obtained by dip-coating subsequently nano-dispersion 2, a colloidal sol and a polymeric sol.
(Nano-dispersion 2, Colloidal sol and Polymeric Sol: Layers made by 2 x dip-coating and calcination at 500°C; (a,b) bar = 1 µm)

Pore analysis of the toplayer material with N_2-adsorption/ desorption measurements indicated a microporous structure (type I isotherm) with a pore diameter of ~ 1 nm (Figure 12a). Further, Figure 12b shows that the cubic polymorph of zirconia was also found for the membrane material made from the nano-structured sol.

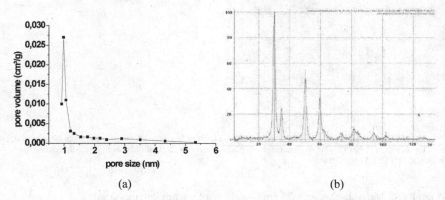

(a)   (b)

Fig. 12. Pore size distribution (a) and phase structure (b) of the polymeric sol-gel derived material after calcination at 500°C.

Figure 13a shows cross-section and surface micrographs of the graded membrane after further firing at 1200°C. From the detail cross-section micrographs shown in Figure 13a (a) and 13a (f), it appears also that a partly sintered YSZ layer was obtained from the nano-dispersion derived layer and that the sol-gel derived layers sintered to full density. The image showed however no distinction between the different sol-gel layers, which suggests that both layers sintered into a single dense layer. Further, it appears that the thickness of the sintered layer can vary significantly in different cross-section pictures; in an area where the surface shows a strong curvature, we found a thickness of several hundred nanometers (13a (f)). Based on SEM micrographs of the surface (Figures 13a (b), 13a (d), 13a (e)), it appears also that after firing at 1200°C completely sintered layers were obtained. For the sintering experiments at 1300°C and 1400°C, the same tendencies were found as previously described.

5. He leak tests

Currently, the densification degree of the different 8YSZ membranes described in this work is further investigated in He leak tests. In Table 1, the results of a first series of coating experiments are summarized (75 x 75 mm² plates, sintering temperature 1400°C). Membrane system 1, prepared by coating nano-dispersion 1 showed the highest He leak rate (1 plate prepared and tested). Membrane system 2 was prepared as an alternative for 1 and showed an improved leak rate (4 plates tested), which can be explained by the significantly smaller particle and pore size of the layers in the calcined state. For the next membrane system which was made by combining the same nano-dispersion coating as in membrane system 2 and a colloidal sol-gel coating, the result was again a slightly improved leak rate (6 plates tested). Then, as could be expected, the lowest leak rates were found for the membrane system with an additional sol-gel layer (nano-dispersion 2, colloidal sol, polymeric sol; 4 plates tested) and for the membrane system with a perfectly graded structure (nano-dispersion 1, nano-dispersion 2, colloidal sol, polymeric sol; 1 plate tested), but it should be mentioned that only a limited amount of test results are available for these membranes.

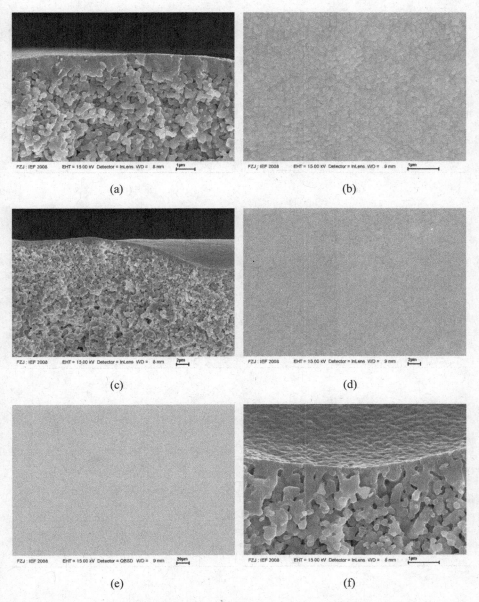

Fig. 13a (a) – (f). Micrographs of the cross-section and surface of the membrane shown in Fig. 11 after firing at 1200°C.
(Nano-dispersion 2, Colloidal sol, Polymeric sol: 2 x dip-coating and calcination at 500°C; (a,b) bar = 1 µm, (c,d) bar = 2 µm, (e) bar = 20 µm, (f) bar = 1 µm)

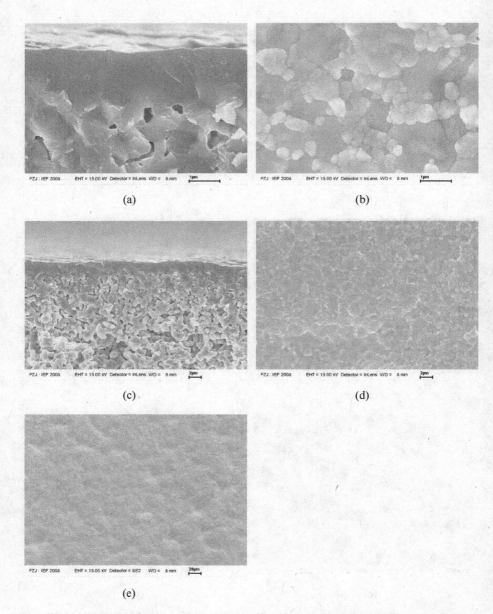

(a)

(b)

(c)

(d)

(e)

Fig. 13b (a) – (e). Micrographs of the cross-section and surface of the membrane shown in Fig. 11 after firing at 1300°C.
(Nano-dispersion 2, Colloidal sol, Polymeric sol: 2 x dip-coating and calcination at 500°C; (a,b) bar = 1 µm, (c,d) bar = 2 µm, (e) bar = 20 µm)

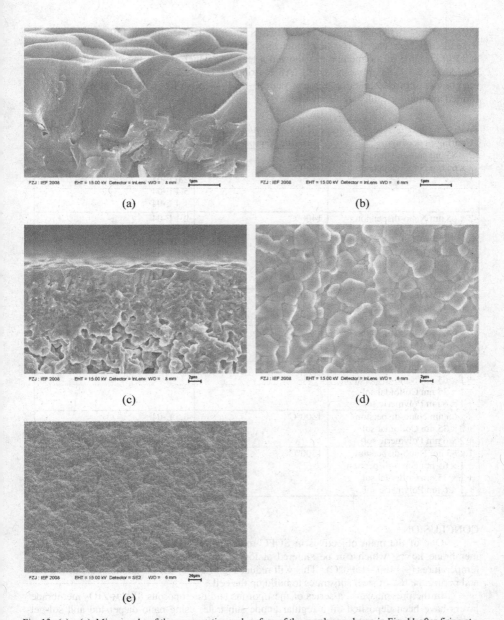

Fig. 13c (a) – (e). Micrographs of the cross-section and surface of the membrane shown in Fig. 11 after firing at 1400°C.
(Nano-dispersion 2, Colloidal sol, Polymeric sol: 2 x dip-coating and calcination at 500°C; (a,b) bar = 1 µm, (c,d) bar = 2 µm, (e) bar = 20 µm)

Also, a first series of membranes with a sintering temperature of 1300°C and 1200°C have already been tested. The first conclusions from these tests were as follows. After firing at 1200°C, none of the membranes showed a leak rate which was in the measuring range of our device ($\sim$ E-03 mbar.l.s^{-1}.cm^{-2}). After firing at 1300°C, on the other hand, the first membranes showed a measurable leak rate of $\sim$ 5 E-04 mbar.l.s^{-1}.cm^{-2}, but also here only a limited amount of results are yet available.

Table 1. He leak test results of sintered 8YSZ thin-film electrolyte membranes, made by spin-coating

Coating procedure	Sintering Temperature	He leak rate (mbar.l.s^{-1}.cm^{-2})
2 x 85 nm Nano-dispersion	1400°C	8 E-04
2 x 65 nm Nano-dispersion	1400°C	6.2 E-04 2.0 E-04 5.4 E-04 1.4 E-04
2 x 65 nm Nano-dispersion + 2 x 35 nm Colloidal sol	1400°C	2.1 E-04 3.1 E-04 6.7 E-05 9.7 E-05 9.3 E-05 5.9 E-05
2 x 65 nm Nano-dispersion + 2 x 35 nm Colloidal sol + 2 x 6 nm Polymeric Sol	1400°C	4.8 E-05 1.4 E-04
2 x 65 nm Nano-dispersion + 1 x 35 nm Colloidal sol + 1 x 6 nm Polymeric sol	1400°C	1,9 E-04 1,9 E-04
1 x 85 nm Nano-dispersion + 1 x 65 nm Nano-dispersion + 1 x 35 nm Colloidal sol + 1 x 6 nm Polymeric sol	1400°C	1.4 E-05
2 x 65 nm Nano-dispersion + 2 x 35 nm Colloidal sol + 2 x 6 nm Polymeric sol	1300°C	6,5 E-04
1 x 85 nm Nano-dispersion + 1 x 65 nm Nano-dispersion + 1 x 35 nm Colloidal sol + 1 x 6 nm Polymeric sol	1300°C	4,9 E-04

CONCLUSION

One of the main objectives in SOFC manufacturing is to produce novel electrolyte membrane layers, which can be sintered at lower temperatures than the currently applied temperatures ($\sim$ 1400 - 1500°C). This will reduce significantly the production cost of the cell and permit the use of steel substrates to build up the cell.

In the present work, a series of mesoporous and microporous $8Y_2O_3$-ZrO_2 membrane layers have been deposited on a regular anode substrate, using nano-dispersion and sol-gel coating procedures. As shown in the first part, deposition of continuous membrane layers on a typical macroporous NiO/YSZ substrate requires a coating liquid containing relatively large particles in order to prevent infiltration in the substrate during the deposition process and to increase the allowable layer thickness. Sintering experiments showed however that for complete sintering of such layers – in our experiments made from particles with a size of 85 nm and 65 nm – a comparably high firing temperature of 1400 °C is also required. From this result, the main advantage of the developed deposition process involves the decreased layer thickness in comparison with regular suspension coating processes (thickness of the finished sintered layer is ~ 1 - 2 μm), but in order to obtain a lower thermal treatment further research was devoted to the development of coating liquids with a smaller particle size.

For applications requiring a lower firing temperature, thin and ultra-thin nano-structured $8Y_2O_3$-ZrO_2 membrane layers were made from sols containing particles with a size of ~ 35 nm and 6 nm. The average pore sizes of the nano-structured membrane material measured ~ 3 – 4 nm and 1 nm. These membranes were coated on a previously deposited YSZ layer made from a nano-dispersion and from the first results of our sintering experiments it was shown that such layers can be sintered at 1200°C.

Further studies are currently devoted to preparing the described $8Y_2O_3$-ZrO_2 membranes gas-tight on our standard anode and in our future work we plan to use one of these new $8Y_2O_3$-ZrO_2 membranes as the electrolyte layer of a steel supported SOFC.

REFERENCES
[1] T. Van Gestel, W.A. Meulenberg, M. Bram, D. Stöver, Development of novel microporous ZrO_2 membranes for H_2/CO_2 separation, submitted for publication in the proceedings of the 33th International Conference and Exposition on Advanced Ceramics and Composites, 2009, American Ceramic Society
[2] T. Van Gestel, D. Sebold, W.A. Meulenberg, H.-P. Buchkremer, Development of thin-film nanostructured electrolyte layers for application in anode-supported solid oxide fuel cells, Solid State Ionics 179 (2008) 428-437
[3] T. Van Gestel, D. Sebold, H. Kruidhof, H.J.M. Bouwmeester, ZrO_2 and TiO_2 membranes for nanofiltration and pervaporation: Part 2. Development of ZrO_2 and TiO_2 toplayers for pervaporation, Journal of Membrane Science 318 (2008) 413-421

ADVANCED CELL DEVELOPMENT FOR INCREASED DIRECT JP-8 PERFORMANCE IN THE LIQUID TIN ANODE SOFC

M. T. Koslowske, W. A. McPhee, L. S. Bateman, M. J. Slaney, J. Bentley and T. T. Tao,
CellTech Power, LLC; Westborough, Massachusetts 01581, USA

ABSTRACT

CellTech Power's direct conversion fuel cell technology based on the Liquid Tin Anode-Solid Oxide Fuel Cell (LTA-SOFC) has been demonstrated to operate on gaseous, solid and liquid carbonaceous fuels without fuel processing, reforming or sulfur removal. For these complex fuels, the LTA-SOFC total system efficiency has been predicted to be >30% for direct JP-8 for portable power applications up to 61% for large coal systems, in addition to being dramatically simpler than other fuel cell systems. Current technology development is directed at using liquid, JP-8, and solid, coal, fuels. In previous programs, the cell power density was increased from 40 mW/cm^2 to 120 mW/cm^2 on direct JP-8 fuel, coupled with a four fold reduction in weight and volume. These improvements resulted in the Gen 3.1 cell design. An ONR program directed towards further advanced cell development will continue the focus on cell component optimization for the target areas of weight reduction (1.5X), performance increase (1.5X) and stability (>100 thermal cycles, >1,000 hrs). Progress will be reported on near term results towards these goals.

INTRODUCTION

The Liquid Tin Anode - Solid Oxide Fuel Cell (LTA-SOFC) has been developed to take advantage of the unique physical and chemical properties of tin. The operational theory of the LTA-SOFC has been described in detail elsewhere [1-4]. The use of a liquid tin anode allows CellTech Power systems to be fuel flexible without auxiliary fuel processing to reform and de-sulfurize heavy fuels. It is this aspect which makes LTA-SOFC attractive for generating power from common fuels such as natural gas, propane, diesel, kerosene, gasoline, biodiesel, ethanol, Fischer-Tropes fuel, biomass and coal [2]. The practical applications of the LTA-SOFC have been demonstrated on a multitude of these carbonaceous fuels at a single cell, stack and system level over the history of the company [3,4]. Two prototype standalone systems rated at 1 kW were demonstrated for two thousand plus hours each on natural gas and hydrogen. It is important to acknowledge that the LTA-SOFC can operate with high efficiency on hydrogen and is well suited as a flexible power generation device when the hydrogen economy arrives. Currently, LTA-SOFC system development is targeted for two areas. The first is portable power applications using liquid petroleum based logistic fuels. The second is land based power generation with coal. This paper will present developments towards portable power systems using JP-8 logistic fuel.

Portable Power

Liquid petroleum based logistic fuels have a significant volumetric and gravimetric energy density advantage compared to other fuels which makes them highly desirable for mobile/portable applications. The US Department of Defense is the largest single consumer of oil in the world, with the majority of this oil being consumed as jet fuel, designated JP-8 [5]. The DOD operates under Directive 4140.25 which gives precedence to the use of JP-8 fuel for all military forward deployed equipment [6]. Increasing deployment of electronic devices on the battlefield has pushed the US military portable power demands to an unprecedented level. A recharging base power supply compatible with the current logistic fuel is desired to simplify battlefield supply chain management. Fuel cells in general offer significant technological advantages for portable power application, with efficiency and reliability quoted as the most advantageous qualities. However, JP-8 has a high sulfur content which can range as high as 3,000 ppm and is severely detrimental to the catalytic activity of conventional fuel cell anodes. The tin-sulfur phase diagram at 1,000°C shows reasonable sulfur solubility in tin, up to 2.5 wt%, with a mixture of liquid tin and tin-sulfide existing up to 20 wt% sulfur [7]. High sulfur concentrations have been tested with little degradation to the tin anode, eliminating the need for reforming and desulfurization [3].

Coal Base Power
Coal provides over half of the base load power generation requirement in the U.S. Recent developments, as a result of high petroleum based energy prices, has shifted focus towards clean and efficient energy production. Clean coal and renewable green energy have highlighted the environmental and efficiency liability of the current coal power generation technology. Internal single cell testing results have shown that efficiency values can be high for the LTA-SOFC. Tests using bio-char (U of Hawaii) and coal (pulverized East/West blend) showed efficiencies of 67 and >57% respectively, after correction for resistance and air leakage. These results are confirmed by an Electronic Power Research Institute report on the Systems Assessment of Direct Carbon Fuel Cell Technology [8]. The results are shown in Figure 1 and include 100% CO_2 capture.

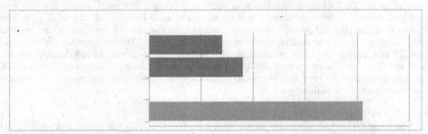

Figure 1. Coal Plant Efficiency Comparison Between Current Technology, Advanced Integrated Gasification Combined Cycle Technology and the Liquid Tin Direct Carbon Concept [8].

Portable Power Challenges
The capability of the LTA-SOFC to operate directly on JP-8 creates a simplicity and efficiency that offers a competitive solution to military and commercial portable power. However, final military specifications are not yet published for this type of device. In order to address the future operational requirements, material development and cell improvement activities have been focused on four key areas:
1. Cell design to increase system level power density.
2. Thermal cycle life and performance consistency.
3. JP-8 only testing.
4. Fast start-up time for power production.

Cell Design
Previously, material developments on the porous separator and cathode current collection have been presented [1]. The porous separator contains the liquid tin on the outside of the electrolyte and allows fuel to react with the tin surface. Modifications of the cathode side current collection resulted in current path uniformity along the active length. These improvements resulted in the Gen 3.1 series of LTA-SOFC designs. Current cell improvements are ongoing and are shown in Figure 2. The Gen 3.1 design resulted in dramatic weight and volume reductions, while increasing cell performance to 100 mW/cm² on JP-8.

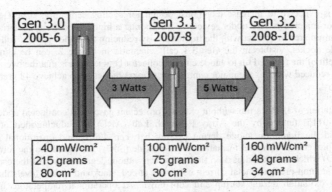

Figure 2. Comparison of CellTech LTA-SOFC physical attributes and cell
performance goals on JP-8

The current cell improvement program is targeted to produce a high performing cell with robust characteristics. The sponsor for the cell development is the Office of Naval Research (ONR). Compared to the Gen 3.1 cell design, the Gen 3.2 will have significant material modifications in order to achieve the aggressive goals of reducing weight and increasing performance by approximately 50%, with the same form factor. Table I lists the performance (power density, cell power), weight (grams, specific power) and operational requirements (lifetime, thermal cycles) for the Gen 3.2 design compared to the Gen 3.1 cell. The primary focus of the material innovations will be a transition to a cathode supported electrolyte component and the anode current collection composition. The

Table I. Gen 3.2 performance goals

ONR GEN 3.2		
PROJECT MILESTONES	CURRENT (GEN 3.1)	2010
JP-8 Power Density (mW/cm2)	100	160
Single Cell Power- JP-8 (W)	3	4.8
Cell Weight (g)	75	48
Cell Specific Power (W/g)	0.04	0.1
Lifetime (Hrs)	100	1000
Thermal Cycles	10	100

development of a single component which incorporates the cathode layer as a support structure to a thin electrolyte layer is not new to high temperature fuel cells, both planar and tubular. The targeted reduction in electrolyte thickness from 180-200 microns for the Gen 3.1 design to between 50 and 75 microns for the Gen 3.2 is expected to result in a significant performance increase. The support of the cathode layer is expected to increase the robust design for cycling capability. The Gen 3.2 anode current collector material is also under review to incorporate a high conductivity material which will reduce the required area in contact with the liquid tin. By reducing the anode current collector contact area, the anode pocket, visible in the Gen 3.1 cell schematic in Figure 2, can be eliminated. An additional benefit of the reduced tin to anode current collector contact area is that the overall amount of tin can then be reduced which is the major contributor to mass reduction to achieve 48 grams.

JP-8 Startup

Investigation of JP-8 startup without external protectant gases was conducted under Contract # W911NF-08-1-0115, funded by the Army Research Lab (ARL). A reducing/inert atmosphere is required around the liquid tin anode during cycling between 400°C and operational temperature, 1,000°C. This prevents detrimental formation of tin dioxide (SnO_2) as opposed to the formation of tin monoxide (SnO) which is reduced back to tin by the fuel without degrading any of the components. In addition, there are non-precious metal current collectors at cell, stack and system levels that need to be protected from oxidation during startup and cool down. At operational temperature fuel (JP-8) is metered into the anode to reduce the tin monoxide (SnO) from the electrochemical reaction back to tin. A practical system will require JP-8 to provide a reducing atmosphere at the anode during thermal cycling. Thermal decomposition of JP-8 at elevated temperatures (>100°C) will give a chemically reducing environment, initially from the light volatile fraction and then from the heavy pyrolization products (soot) reacting with oxygen to form carbon monoxide or carbon dioxide. A solid fuel such as carbon can be used to reduce tin oxide back to tin via one of the following pathways. Equations (1) and (2) describe direct oxidation of carbon, which has been demonstrated by CellTech Power. The Generation 3.1 cell incorporates a porous ceramic which provides containment of the tin anode and allows only gaseous products to interact with the tin. In this case, Eqs (3) and (4) describe the oxidation process. Analysis of Equation (4) using HCS Chemistry software version 4.1, indicates that

$$SnO_2{}_{(s)} + 2 C_{(s)} = 2 CO_{(g)} + Sn_{(l)} \tag{1}$$

$$SnO_2{}_{(s)} + C_{(s)} = 2 CO_2{}_{(g)} + Sn_{(l)} \tag{2}$$

$$SnO_2{}_{(s)} + CO_{(g)} = CO_2{}_{(g)} + Sn_{(l)} \tag{3}$$

$$CO_2{}_{(g)} + C_{(s)} = 2 CO_{(g)} \tag{4}$$

the reaction is spontaneous above 700°C. Figure 3 shows the temperature dependence of ΔG for equation (4) to 1,000°C where the ΔG is -52.2 kJ given one mole of reactants.

The LTA-SOFC system can consume excess carbon deposited by the JP-8 during heating and cooling to maintain a reducing atmosphere. However, this reducing atmosphere has to be managed to ensure all components susceptible to oxidation at elevated temperatures are protected. In addition the metering of the fuel into the anode chamber has to be controlled during operation so that there is no excessive carbon build up (soot). The presence of soot can be detrimental to the operation of the device, such as shorting current paths or blocking fuel diffusion through the porous separator.

Fast Heating

The requirement of fast heating and cooling cannot be demonstrated using the typical resistance heating element furnaces such as the ones used for single cell testing. Furnace modifications resulted in heating from room temperature to 1,000°C in <30 minutes. However, cooling is typically slower than heating. According to a draft battery charging specification, the requirement for startup is operation within 10 minutes [9]. The LTA-SOFC is able to provide a minimal power profile as low as 800°C, which can be used to assist the heating of the cell or provide external power.

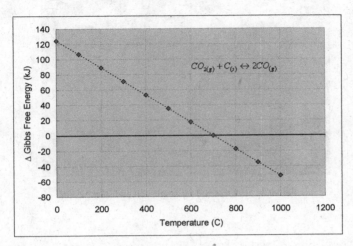

$$CO_{2(g)} + C_{(s)} \leftrightarrow 2CO_{(g)}$$

Figure 3. The reaction of solid carbon with carbon dioxide to form carbon monoxide is spontaneous as represented by the negative change in Gibbs free energy for the reaction.

EXPERIMENTAL

The single cell testing procedure has been detailed in a previous paper [1]. Cell performance and component testing has continued with the Gen 3.1 cell design. The cell components have been under testing for the thermal cycling capability using both hydrogen and JP-8 as fuel sources. In addition, component evaluation of the porous separator thickness was performed in a JP-8 environment. Performance testing using JP-8 consists of heating to temperature under a hydrogen (3% H_2O) atmosphere, then switching to JP-8 for the testing. Thermal cycle testing with JP-8 utilized the fuel between 400°C – 1,000°C. JP-8 flow rate for all testing presented in this paper was 50 μl/min. The Gen 3.2 cell testing was not completed at the time of this paper.

An independent combustion device that runs on JP-8 was used to fabricate a stand alone JP-8 test bed. Figure 4 shows the test bed setup. The JP-8 is combusted in an interior chamber which heats the cell enclosure by radiation. The cell enclosure is sealed from the combustion gases and a separate JP-8 injection device is used to provide fuel to the test articles. By keeping the combustion products from the fuel cell chamber, the fuel rates can be precisely controlled for performance testing and optimization. This system is capable of cold startup, cooling and cycling solely on JP-8 as heat source and fuel source.

Figure 4. JP-8 test bed for single cell and stack testing

RESULTS

Cell Design

Weight reduction is a critical metric for portable power devices. The new Gen 3.2 cathode supported electrolyte is shown in Figure 5. The mass of the integrated component is roughly the same as the Gen 3.1 electrolyte assembly with the equivalent construction. The average is 17.6 grams. Though there is no significant mass reduction at this time, the difference in electrolyte thickness is expected to have a dramatic effect on cell performance.

Figure 5. Comparison of Gen 3.1 (top) to Gen 3.2 (bottom) electrolytes.

The porous separator is a candidate component for weight reduction by decreasing the wall thickness. Table II shows the relationship between wall thickness and total component mass. Currently the nominal wall thickness for the Gen 3.1 test assembly is 2.5 mm. A reduction to 1.5 mm will result in a weight savings of 8.2 grams per cell or ~30% of the required reduction. The effect the wall thickness on JP-8 performance was tested. Figure 6 shows the V-I curve performance with a constant flow rate of JP-8 for the various porous separator thicknesses.

Table II. Comparison of porous separator mass to finished wall thickness.

Wall thickness	(mm)	3	2.5	2	1.5
Component mass	(g)	27.8	19.8	16.1	11.6

The performance is nearly equivalent, with the 2.0 and 1.5 mm samples showing higher current before the cell becomes unstable. One explanation is that the diffusion through the porous separator is faster in the thinner material. A modeling program is underway with the University of South Carolina to create a comprehensive cell model for the LTA-SOFC.

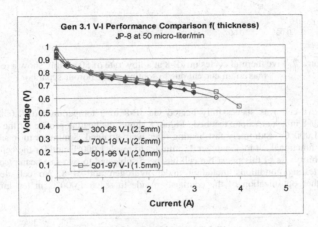

Figure 6. Porous separator thickness comparison of JP-8 performance.

Thermal cycling

Several cells were thermally cycled on hydrogen and JP-8. The Gen 3.1 cell completed 13 thermal cycles in 2007. Currently a Gen 3.1 cell has completed more than 20 thermal cycles with a hydrogen fuel source. This cell will be cycled to material failure for use in developing components for the Gen 3.2 cell requirement of 100 thermal cycles. Under JP-8 startup and cycle testing, a Gen 3.1 cell was cycled five times. Figure 7 presents the JP-8 performance V-I-W curves. The test resulted in little performance degradation as shown by the overlapping slopes. The later cycles were not able to maintain high performance due to excess carbon build up, which effectively closed off the porous separator to high fuel diffusion to the tin surface and reactant product exchange away from the tin surface.

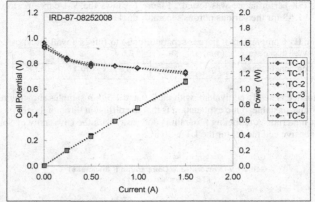

Figure 7. Five thermal cycles on JP-8 at a flow rate of 50 μl/min, show a reduction in maximum current due to sooting of the porous separator

Fast Heatup

A Gen 3.1 cell was successfully heated via JP-8 to a temperature of 1,000°C and tested completely on JP-8 in a stand alone test bed. Figure 8 shows the heating profile for the test from room temperature to 1,000°C. Extremely fast heating rates were attained, 8 minutes to reach 800°C, when the cell became functional. It took another 15 minutes to reach 1,000°C and for the cell to be tested to full power capability. As of this paper the cell has been subjected to two fast heatup/cooldown cycles under various testing conditions including under hydrogen and JP-8. No cell degradation was observed. Further optimization of the combustion side to reach 1,000°C in ten minutes is under consideration.

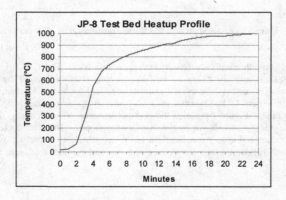

Figure 8. Heating profile for JP-8 test bed

CONCLUSIONS
The material develop to increase cell performance is continuing with the development of the Gen 3.2 cathode supported electrolyte cell. The component has been fabricated, but unfortunately not tested at the time of this paper. Advancements have been made with the use of JP-8 to bring the LTA-SOFC technology closer to a reality for portable power applications.

- A Gen 3.1 cell has been thermally cycled for >20 times.
- Thermal cycling of a Gen3.1 cell using only JP-8 resulted in 5 thermal cycles before excess sooting/carbon formation resulted in decreased fuel diffusion. The current collection was also protected.
- Porous separator thickness does not dramatically effect JP-8 cell performance, however thinner wall thickness has a positive impact on weight reduction.
- Fast heatup times, eight minutes to 800°C, were achieved on a JP-8 test bed.

ACKNOWLEDGEMENTS
Development of Gen 3.0 was partially supported under DARPA Contract W911QY-04-2-003;
Development of Gen 3.1 was partially supported under DARPA/ARMY Contract W911NF-07-C-0032
Development of JP-8 startup was partially supported under ARL Contract W911NF-08-1-0115

REFERENCES
[1] M.T.Koslowske et al., Performance of the Gen 3.1 Liquid Tin Anode SOFC on Direct JP-8 Fuel, *Advances in Solid Oxide Fuel Cells IV, CESP*, **29**, 5, 39-50 (2008)
[2] J. Bentley, T, Tao, Liquid Tin Direct Fuel Cell for JP-8, Coal and Biomass Applications, *DOD 6th Annual Logistic Fuel Processing Conference*, May 2006.
[3] T. Tao, L. Bateman, J. Bentley and M. Slaney, Liquid Tin Anode Solid Oxide Fuel Cell for Direct Carbonaceous Fuel Conversion, *Fuel Cell Seminar*, 2006, p. 198
[4] T. Tao, Introduction of Liquid Anode/Solid Oxide Electrolyte Fuel Cell and its Direct Energy Conversion using Waste Plastics, in *SOFC-IX*, S.C. Singhal and J. Mizusaki, editor, March 2005, Vol 1, p 353
[5] S. Karbuz, Energy Bulletin, Feb (2007)
[6] Wolfowitz, P., 2004, *Department of Defense Directive 4140.25 – DoD Management Policy for Energy Commodities and Related Services*, Department of Defense, Washington, pp.1-10.
[7] R. S. Roth, T. Negas, and L. P. Cook, Editors, Phase Diagrams for Ceramists, Vol V, The American Ceramic Society, (USA).
[8] Systems Assessment of Direct Carbon Fuel Cell Technology, EPRI, Palo Alto, CA: 2008. 1016170.
[9] Draft Fuel Cell Battery Charger System Specification 250-500 W Nominal, US Army Research Development and Engineering Command, Communications-Electronics Research, Development, and Engineering Center, 29 SEPT 2004

FUNDAMENTALS OF LIQUID TIN ANODE SOLID OXIDE FUEL CELL (LTA-SOFC) OPERATION

Randall Gemmen[1], Harry Abernathy[1], Kirk Gerdes[1], Mark Koslowske[2], William A. McPhee[2], Tomas Tao[2]

[1]National Energy Technology Laboratory, Morgantown, WV
[2]CellTech Power, LLC, 131 Flanders Rd., Westborough, MA 01581

ABSTRACT

An alternative high temperature fuel cell system, called Liquid Tin Anode Solid Oxide Fuel Cell (LTA-SOFC) technology, is presently under consideration by NETL given its ability to directly convert coal. Before such a fuel cell concept can be considered in system studies for commercial development, a detailed assessment for the electrochemical activity and oxygen diffusion within the liquid tin is needed. In addition, the fundamental thermodynamic operation of such a concept needs to be properly analyzed. In the present work, initial research efforts to characterize the tin electrochemistry on a button cell at 900°C and with a tin thickness of 6mm show that both the temperature and the fuel type control the EIS signatures, and, hence, the various overpotentials present. Preliminary analysis of the data also suggests an *effective* oxygen diffusion coefficient of about 1E-3 cm^2/s for the conditions tested, giving evidence that the process is not purely diffusional through the anode. A detailed model that includes the multiple coupled reaction and transport mechanisms present will be required in order to describe oxygen transport in the tin.

1.0 INTRODUCTION

According to the International Energy Agency, world-wide demand for energy will continue to expand at rates of about 1.6% per year[1], which by 2030 will accrue to an overall 45% increase from 2006. Much of this energy will be electric-based, especially if plug-in hybrid vehicles become the favored mode of transportation. Even with the expected high growth of renewable energy technology, less than 5% of total power generation is expected to come from non-hydro renewables by 2030[1]. Hence, it is clear that for purposes of achieving long term energy security, electric power production from coal can be expected for years to come. To minimize the impact to the environment, CO_2 emissions from coal plants will need to be captured and sequestered, and much work is being done to develop and achieve such capability[2,3].

To minimize the cost and environmental impact of electric energy production from coal, high efficiency systems are required. High temperature fuel cells, such as solid oxide fuel cells (SOFC), efficiently convert fuel energy in to electricity through electrochemical reactions (vs. combustion reactions that produce only heat.) This technology can convert the chemical energy from a wide variety of fuel sources such as coal, biomass, and various hydrocarbon waste streams such as wood, paper, and plastics, into electricity[4]. An important consideration for any of these fuels is managing the impact of trace elements contained in these fuels on fuel cell component degradation. At the high temperatures of operation of an SOFC (ca. 750 to 1000 °C) where good fuel conversion kinetics can be achieved, these trace fuel constituents react with the cell materials in the anode thereby reducing the electrochemical reaction rates, increasing the electric ohmic resistances of their materials, and causing material phase changes that can weaken materials and result in mechanical failure.

Mitigating the impact of such reactions requires reduction of the trace element loading via filters and sorbent catalysts, or increasing the robustness of the anode materials to improve tolerance for exposure to contaminant materials. In this paper we consider a metal anode material that resides as a liquid layer between the fuel gas and the cell electrolyte layer, Figure 1. An example of such technology is CellTech Power's direct conversion fuel cell technology based on the Liquid Tin Anode-Solid Oxide Fuel Cell (LTA-SOFC). This technology has been demonstrated to operate on gaseous, solid and liquid carbonaceous fuels without fuel processing, reforming or sulfur removal[4]. System efficiency is expected to be 61% for coal based systems. Current state of the art single cell performance shows a power density of >160 mW/cm² on hydrogen and 120 mW/cm² on direct liquid JP-8 fuel.

As shown in Figure 1, since the anode is a liquid layer it fully covers the electrolyte active domain where oxygen is exchanged between the electrolyte and the anode. Hence, it can serve as a blocking medium for contaminants in the gas and inhibit their transport to the electrolyte. In addition, for fuel contaminants that are possibly dissolvable within the liquid layer, a transport-resistance layer is offered which can impede the rate of contaminant reaction with the electrolyte thereby slowing the rate of degradation. Fuel contaminants that can be electrochemically oxidized could also provide a fuel source[5]. Finally, while meager data exist, it can be expected that because the liquid layer fully covers the electrolyte, the "electrolyte surface usage efficiency," η_{es} is improved over existing porous solid anode technology which at best utilizes only the electrolyte surface located near the so-called 'triple-phase-boundary'. Hence, oxygen reactions can be expected to occur over the full surface of the electrolyte when using a liquid anode.

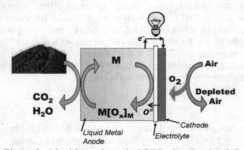

Figure 1. Liquid metal anode SOFC. See Section 2 for Nomenclature.

In spite of the above potential advantages of using liquid anodes, several factors need to be considered before a significant effort is made to develop these commercially. Foremost is developing an understanding of how oxygen and fuel react within the liquid metal electrochemical system. In particular, while the high interfacial contact, η_{es}, between the anode and the electrolyte is certainly beneficial, the electrochemical oxidation kinetics can be relatively slow for liquid metal anodes so as to offset such benefits. Details of the kinetics of the oxygen transfer at the liquid metal – electrolyte interface need to be understood. Also, oxygen transport within the liquid anode needs to be understood. This problem is complicated by the unknown in-situ oxygen state—whether it occurs as a metal oxide, metal suboxide, or dissolved oxygen specie. This paper will examine some of the presently available data for tin-based SOFCs, and describe our strategy to accurately measure the relevant parameters.

Section 2 of this paper presents some of the thermodynamic considerations for the operation of a liquid metal anode SOFC. Because the metal anode simply acts as an intermediary in the overall thermodynamic state change for fuel combustion, these results are general to any liquid metal anode SOFC. Special consideration will also be given to when the system acts as a battery (i.e., converts the metal to a metal oxide due to lack of sufficient fuel). Section 3 focuses the discussion on tin for the liquid metal anode by presenting available property data for tin. Section 4 describes possible pathways for oxygen transport within the metal anode, with focus on tin. Section 5 shows the experimental approach being used to study two controlling mechanisms, diffusion transport and reaction kinetics, and presents available experimental data. Section 6 provides a summary of the work and reviews future work.

2.0 THERMODYNAMICS OF LIQUID ANODE SOFC OPERATION

In the operation of a liquid metal-SOFC system (fuel cell mode), the liquid anode will participate as an 'intermediary' for the oxidation of fuel delivered to the fuel cell. As an example, consider a tin-SOFC fueled by hydrogen. This case can be written concisely using the following reaction steps on the anode side:

$$Sn + xO^= \rightarrow Sn[O_x]_{Sn} + 2x\ e^- \tag{2.1}$$

$$xH_2 + Sn[O_x]_{Sn} \rightarrow Sn + xH_2O \tag{2.2}$$

Reaction 2.1 occurs at the anode-electrolyte interface. Reaction 2.2 occurs at the gas-anode interface. In the above equations, $O^=$ is oxygen ion supplied to the anode by the electrolyte. $[O_x]_{Sn}$ is a state of oxygen within the metal anode (tin for the reactions shown in these equations). The use of the parameter 'x' is required, because the state of oxygen in the tin is most likely varied depending on operating conditions. Oxygen may associate tightly with Sn to form an oxide or suboxide, or more loosely as some dissolved oxygen state within the liquid metal. Although several $[O_x]_{Sn}$ states can be considered, the specific form is irrelevant for the present thermodynamic analysis.

Adding the two equations, 2.1 and 2.2, provides the overall change for the reactions on the anode. This overall change on the anode shows the oxidation of hydrogen with O-ion ($H_2 + O^= \rightarrow H_2O + 2e^-$). Such an overall change is the same for any liquid metal based SOFC anode system, albeit the value of 'x' in the detailed reactions, 2.1 and 2.2, may be particular to a given metal-oxide complex. Hence, in principle one can envision a liquid Ni anode with similar functionality. Further, it should be evident that the overall reaction is the same as that occurring for any other SOFC technology (e.g., porous Ni-YSZ based anode technology), and in fact is general to any operating oxygen-ion-conducting fuel cell.

Ideal Potential

Given the above analysis, we can directly conclude that the ideal potential for a liquid metal anode SOFC will be the same as any other fuel cell, and when operating on hydrogen fuel can be written as[6]:

$$E_N = \frac{-\Delta G^\circ}{2F} + \frac{RT}{2F} \ln\left(\frac{a_{H2-A} a_{O2-C}^{1/2}}{a_{H2O-A}} \right) \tag{2.3a}$$

or, more generally as,

$$E_N = \frac{RT}{4F} \ln\left(\frac{a_{O2-C}}{a_{O2-A}}\right) \qquad (2.3b)$$

In Equation 2.3, E_N is the ideal Nernst voltage, ΔG^o is the change in the standard state Gibbs free energy for the overall reaction, $H_2 + \frac{1}{2}O_2 \rightarrow H_2O$, which is only a function of temperature, a_{z-y} is the gas phase activity (partial pressures when assuming ideal gas behavior) of a supplied reactant ('z') present in either the cathode or anode ('y') gas, R is the Universal Gas Constant, T is absolute temperature, and F is the Faraday constant.

Anode Gas – Liquid Metal Equilibrium

Because thermodynamic equilibrium is assumed to exist when analyzing for E_N, anode gas-phase species are in equilibrium with their counterparts in the liquid anode. Such equilibrium occurs when the chemical potential of a species in the gas phase is equal to the chemical potential of the same species in the liquid metal. For oxygen within the anode, we can write at equilibrium[7]:

$$\mu_{O2-A} = \mu_{O2-M}$$
$$\mu_{O2}^o + RT \ln(a_{O2-A}) = \mu_{O2}^o + RT \ln(a_{O2-M}) \qquad (2.4)$$
$$a_{O2-A} = a_{O2-M}$$

where μ_{O2-M} stands for the chemical potential of O_2 in the liquid metal anode, where a_{O2-M} is the activity of O_2 within the liquid metal.

Hence, for a given concentration of H_2 and H_2O present in the anode gas, a given activity of oxygen will exist within the gas phase, and this will control the activity of oxygen within the liquid metal, regardless of the dominant form of oxygen existing within the liquid metal anode (e.g., oxide, suboxide, or dissolved oxygen.) As a result, we can write equivalently for Equation 2.3b:

$$E_N = \frac{RT}{4F} \ln\left(\frac{a_{O2-C}}{a_{O2-M}}\right) \qquad (2.5)$$

In summary, at equilibrium, the activity of oxygen within the tin is uniform everywhere, and equal to the activity of oxygen in the anode gas chamber, and provides an equivalent value for E_N as that given by Equation 2.3b.

Liquid Metal – Metal Oxide Equilibrium

The above equations are general, and will apply at any concentration of oxygen within the anode. At sufficiently large concentration of oxygen in the anode [ca. >0.5% O in Sn], a metal-oxide phase can form and equilibrate with the liquid metal and oxygen present in the anode. As shown above, (for thermodynamic analysis) the oxygen in the gas is in equilibrium with the oxygen in the liquid metal, and hence the source of oxygen in the reaction equation can be considered as from either the gas or the metal:

$$M + (x/2)O_2 \rightarrow MO_x \qquad (2.6)$$

At equilibrium, the change in Gibbs free energy is zero:

$$\Delta G = 0 = \Delta G^{o}{}_{M-MOx} - RT \ln(a_{O2-A}{}^{1/2}) \tag{2.7}$$

Solving for the activity of molecular oxygen in the anode, we have:

$$a_{O2-A}{}^{1/2} = \exp(\frac{\Delta G^{o}{}_{M-MOx}}{RT}) \tag{2.8}$$

Hence, at a point where both metal and metal-oxide phases are present within the liquid metal, a specific activity of oxygen within the anode gas is expected (for a specific operating temperature) as shown in Equation 2.8. From Equation 2.3b it can be concluded that until enough oxygen is added to the anode so that all metal is consumed (or enough oxygen is removed from the anode so that no metal oxide is present), the ideal potential will only be a function of the cathode oxygen activity and operating temperature:

$$E_{N} = \frac{-\Delta G^{o}{}_{M-MOx}}{2F} + \frac{RT}{4F} \ln(a_{O2-C}) \tag{2.9}$$

Summary and Estimates for Non-Equilibrium Operation

In summary, the general equation for determining the ideal potential is given by Equation 2.3b. If it is known that both metal and metal-oxide phases are present (oxygen activity at or above the solubility limit), then the Nernst equation, 2.9, can be used (e.g., in battery mode where only M to MO_x conversion is possible). Alternatively, if H_2 is available for reaction, then the Nernst Equation, 2.3, for H_2 to H_2O conversion can be used. As an example, Figure 2 shows both equations plotted over a range of temperatures for a tin-SOFC system, where a specific hydrogen content (97%) is assumed for the use of Equation 2.3. For such high hydrogen concentrations, the gas phase anode oxygen mole fraction is very low, ca. 2.7E-18 at T = 1000 °C, and the tin is nearly completely reduced.

Figure 3 shows for fixed temperature, how oxygen content (O_2/Sn ratio) in the anode controls the ideal potential. For the 1000 °C case, below about 0.25% O_2:Sn the oxygen activity increases with increase in oxygen content until the gas phase oxygen mole fraction is ca. 8.8E-14. As oxygen is added to this system, the oxygen activity within the metal anode becomes fixed until all Sn is converted to SnO_2 (ca. O_2:Sn = 1.0). During this process, the Nernst potential remains fixed at 0.8 V. For oxygen content of O_2:Sn > 1, the anode oxygen activity will rapidly increase and the ideal potential will decrease.

While the above theoretical equilibrium analysis is helpful for estimating fuel cell performance, in a real operating fuel cell (i>0.0 A/cm^2), the system is not in equilibrium, and a higher activity of oxygen will exist within the liquid metal anode than that within the anode gas stream. Specifically, the activity of oxygen near the liquid-electrolyte interface will be greater than that near the gas-liquid interface due to diffusion resistance through the liquid metal. At some level of current density, interfacial oxygen activity can increase sufficiently above that in the anode gas (which can be maintained low in the presence of fuel), and a metal oxide, e.g., SnO_2, will form. As most metal oxides have high melting temperatures, precipitates at the electrolyte interface can form thereby causing a significant reduction in the operating potential.

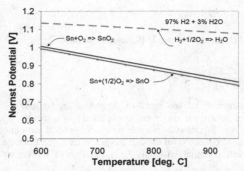

Figure 2. Nernst potentials for $Sn + O_2 \rightarrow SnO_2$, $Sn + (1/2)O_2 \rightarrow SnO$, and $H_2 + \frac{1}{2}O_2 \rightarrow H_2O$. Pressure = 1 atm. Cathode air O_2 partial pressure = 0.21 atm.

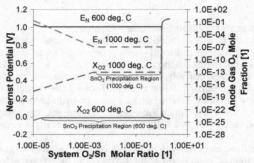

Figure 3. Nernst potentials and anode oxygen mole fractions for a tin + oxygen system with varying anode oxygen:tin molar content, cathode oxygen mole fraction = 0.21, P = 1 atm., and two different temperatures. Concentration data for oxygen in tin calculated using FactSage software.

Finally, depending on the electrolyte, the oxygen transference number may be less than unity which can cause a build-up of oxygen content within the anode liquid metal (especially given the low oxygen diffusion within the liquid metal), and result in open circuit voltages (OCV) lower than commonly seen in porous nickel anodes. Hence, the two potentials given in Figure 2 can be considered as bounds on actual operating OCV's of a tin fuel cell.

3.0 TIN THERMOCHEMICAL PROPERTY DATA

The foregoing discussion has been generalized for any liquid metal anode system, with some specific cases given on tin. Since this work is focusing on the use of tin for the anode, available property data for tin are given here. Although tin has been extensively studied regarding its solid phase properties, and has been used in many applications such as containers, wrappings, coatings and claddings, the specific liquid phase material properties pertinent to this electrochemical application have not been well studied. For example, a wide range in values exist for oxygen solubility in liquid tin.

Basic Property Data

Tin is a potential liquid anode material for SOFCs because of its combination of low melting temperature (~231.88 °C) [8] and high boiling temperature (2270 °C[8,9] to 2603 °C[10]). The variation in boiling point values can be attributed to the discrepancies between experimental data (2270°C) and thermodynamically calculated values by Cahen et al.[10] (2473 °C) and Mcpherson and Hansen (2603 °C) (see Fig. 2 of Cahen et al.). The low melting point allows for improved oxygen solubility by operating far above its fusion point (e.g., 1000 °C). The high boiling point provides low vapor pressures (ca. 0.7 Pa at 1000 °C), and therefore low loss of material when operating at SOFC temperatures. Table I is a compilation of physical property data for tin, including values for oxygen and hydrogen solubility[8]. As will be shown in Section 5, the existence of hydrogen in the tin may explain the faster transient behavior identified in the present tests.

Table I. Physical Property Data for Tin[8]

Property	Temperature (°C)	Value
Fusion Point	-	231.88 °C
Boiling Point	-	2270 °C
Vapor pressure	727	0.001 Pa
	1127	5.9 Pa
Thermal Expansion	0	19.9×10^{-6} ppm/°C
	100	23.8×10^{-6} ppm/°C
Surface Tension	231.88	544 dynes/cm
Viscosity	231.88	1.85 mNs/m²
Expansion on Melting	231.88	2.3 %
Gas Solubility (H_2)	1000	0.04 wt %
	1300	0.36 wt %
Gas Solubility (O)	536	0.00018 wt %
	750	0.0049 wt %

Table II. Temperature Dependent Properties of Tin

Temperature °C	Density[11] g/cm³	Surface Tension[11] dynes/cm	Viscosity[11] cP	Resistivity[12] µOhm-cm
13 (alpha)	5.77			
18 (beta)	7.29			
100				15.5
200		685		20.0
232 solid	7.17			22.0
232 liquid	6.97		2.71	45.0
250			1.88	
300	6.92		1.66	46.8
400	6.85	580	1.38	49.0
500	6.78	565	1.18	51.5
600	6.71	550	1.05	54.0
700		535	0.95	56.3
800	6.57	520	0.87	58.7
900				61.2

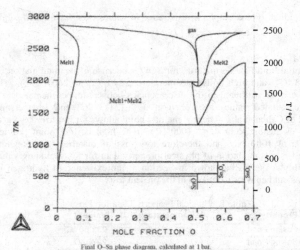

Final O–Sn phase diagram, calculated at 1 bar.

Figure 4. Phase diagram for Sn-O[10] (Celcius scale added).

Other temperature dependant properties that are of interest in the design of high temperature liquid tin solid oxide fuel cells are shown in Table II [9, 11, 12]. There is a phase change from grey (alpha) tin to white (beta) tin at 17°C. As can be seen, there is a need to identify tin data at the high operational temperature and chemical environment of solid oxide fuel cell systems.

Surface tension measurements of drops of liquid tin were also reported by Ricci, et al.[13] as a function of temperature and oxygen partial pressure. Surface tensions between 520 and 550 dynes/cm are reported between temperatures of 352 and 827 °C at low oxygen partial pressure for several heating/cooling rates. In the design of LTA-SOFC systems, surface tension will play a role in how tin contacts the electrolyte.

Sn-O Phase Diagram

The phase diagram for oxygen and tin provides fundamental information on the possible states for oxygen in tin. However, a review of available data shows variability in oxygen solubility in tin. An early version of the Sn-O phase diagram presented by Hansen[14] did not define the oxygen solubility at low concentrations, which was indicated by dashed boundary lines. A more recent version of the phase diagram for the Sn–O system determined by Cahen et al.[10] is given in Figure 4. The thermodynamic model used for the liquid phase of this diagram is Hillert's partially ionic liquid model. The electronic structure of the liquid tin via such a model allows for the solubility of oxygen [O] in the liquid (Melt1) at low concentrations of oxygen. The solubility limit of oxygen is ~2% at 1000 °C. Above this concentration of oxygen, SnO$_2$ begins to precipitate (SnO$_2$ m.p. = 2000 °C [10]).

The solubility of oxygen in tin was also investigated by Bedford and Alcock[15] and is presented in Table III. The values from e.m.f. measurements were only obtained between 500°C and 751°C.

From the experimental data, equation (3.1) can be derived, where N[Sn] is the mole fraction of tin and the term $N^{1/2}$[Sn]c_s corresponds to the equilibrium constant for the reaction ½ SnO_2(s) = ½ Sn(l) + [O]. By fitting equation (3.1) into the form of the error equation (3.2), the two can be used to predict the solubility in the temperature range of 232 to 1100°C. This is possible provided the conditions of oxygen in tin follows Henry's law and the activity of tin obeys Raoult's law for values up to 1.0 at. % [O][15].

$$\log_{10} (N^{1/2}[Sn]c_s) = -5670 / T + 4.12 \tag{3.1}$$

$$\log_{10} (N^{1/2}[Sn]c_s) = -(5730 \pm 210) / T + (4.19 \pm 0.23) \tag{3.2}$$

Experimental work was performed at CellTech Power in an attempt to validate the oxygen solubility in the tin anode fuel cell and to potentially narrow down the range of values found in the literature. CellTech Power Gen 2.0 cells were tested electrochemically to determine the voltage decay at various load values in an inert anode environment. Since the conversion of Sn to SnO_2 is calculated to occur below 0.78V, the [O] can be calculated by the [O$^-$] flux through the electrolyte at each current value. Based on this experimental method, the oxygen solubility was shown to be closer to 0.8 at % at the 1,000°C operating temperature.

Table III. Oxygen Solubility in Tin[15]

Temp. °C	at% [O]	wt% [O]	Method
536	0.0012	0.0002	Measured
600	0.0042	0.0006	Measured
700	0.0190	0.0026	Measured
751	0.0360	0.0048	Measured
800	0.0708	0.0096	Calculated
850	0.1223	0.0165	Calculated
900	0.2019	0.0273	Calculated
950	0.3197	0.0432	Calculated
1000	0.4885	0.0660	Calculated
1050	0.7227	0.0976	Calculated
1100	1.0391	0.1403	Calculated

Finally, the present authors determined the oxygen solubility of the Sn-O system using the thermodynamic modeling software, FactSage version 5.6, based on the available thermodynamic data[16,17,18,19,20]. Figure 5 shows oxygen solubility through a temperature range of -173 < T < 2727 °C, with a maximum of ~10% occurring at 1600 °C. At the fuel cell operating temperature of 1000 °C, the oxygen atom solubility is predicted to be about 0.5%, which is similar to that provided by Belford and Alcock[15], but much different than that shown by Cahen et al.[10]. Differences in the phase diagrams demonstrate the variability in the understanding of the solubility limit of oxygen and other species in tin. Finally, the FactSage analysis does not account for a second liquid phase or intermediate compounds that create a miscibility gap at 1040 °C (as evident in Figure 4). The gas phase regions also show some differences in the maximum oxygen solubility. The shape inconsistency of the Sn-liq + gas boundary can likely be attributed to the assumptions used for the gas phase mixture and the thermodynamic data available in FactSage.

Although the oxygen solubility limit cannot be clearly resolved as yet, several regions within the phase diagrams are pertinent to understanding oxygen transport within the tin-anode SOFC. All data show a lower solubility limit of oxygen in liquid tin for temperatures 477 < T < 1600 °C and oxygen atom concentrations are less than ca. 2% at 1000 °C. Addition of oxygen beyond the solubility limit will precipitate a phase of SnO_2, Sn_xO_{1+x}, or SnO depending on the operating temperature. Differential scanning calorimetry measurements were made by Bonicelli et al., confirming that only SnO_2 is found at temperatures > 525 °C when disproportioning SnO[21].

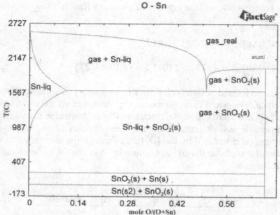

Figure 5. Solubility of oxygen in Sn-O system as a function of temperature.
Total pressure = 1 bar.

As a result of the work of Bonicelli et al.[21] showing SnO_2 present above 525 °C, it is prudent to examine some of the thermochemical properties for SnO_2 systems. Tin dioxide is an n-type wide gap semiconductor with a bulk conductivity on the order of 0.3 S/cm at 1000°C [22]. A chemical diffusion coefficient for oxygen in SnO_2 is reported by Kamp et al.[23]. Through a temperature range of 700 to 1000 °C, the chemical diffusion coefficient is expressed as $D^\delta=0.02 cm^2/s \, exp(-0.9 \, eV/kT)$. Finally, Nielsen et al. report values for the heat of formation of SnO and SnO_2 as 38+/-17 and 42+/-17 kJ/mol, respectively, as determined by ab initio methods[24].

As can be seen, there are still differences in the literature values, both experimental and predicted, for the oxygen solubility in liquid tin. There is a need for a complete scientific understanding of this interaction as it applies to the fuel cell application. Information on the physical properties, such as contact angle and surface tension, for high temperatures and various environments (oxidizing to reducing) have also not been well defined, but are also critical to achieve designs having long term stability. This is critical information for the prediction, modeling and design of real systems.

4.0 LIQUID METAL SOFC OPERATION

Battery Mode

As explained in Section 2, if the metal anode material is readily oxidizable, then it can also act as a source of fuel and can offer another reaction for determining a Nernst potential. If no fuel is present for removing oxygen from the liquid metal (e.g., tin), then the metal-SOFC system becomes a battery whereby the finite quantity of metal present will be steadily oxidized until, essentially all metal becomes metal-oxide (e.g., SnO_2). Over the course of consumption of the tin, the amount of oxygen gradually increases until the tin is saturated with oxygen at an activity equal to that of the cathode. Based on Equation 2.9, this must occur after essentially all metal is oxidized and any excess oxygen exists within the anode gas at an activity equal to the cathode activity. This inherent battery capability provides an extra source of capacity to the system that can buffer fluctuations in fuel gas supply, thereby providing a more stable voltage output.

Operational Losses

In the analysis of common SOFC performance where the anode is supplied gaseous fuel and the cathode is supplied air, the cell voltage at a given current load is often represented as:

$$Vcell = E_N - \eta_{aa} - \eta_{ca} - \eta_{ac} - \eta_{cc} - \eta_R \tag{4.1}$$

where η_i is the loss in voltage due to particular mechanism 'i': aa=anode activation loss; ca=cathode activation loss; ac=anode concentration loss; cc=cathode concentration loss; and R=ohmic loss. These losses will also exist for the liquid anode-SOFC fuel cell. However, notable differences for the concentration loss mechanism can be expected, and new loss mechanisms may also exist. Identifying the existence of new loss mechanisms and devising experiments to characterize them will be necessary.

One additional loss mechanism unique to the liquid anode-SOFC system is the kinetics of fuel oxidation at the anode-gas interface, written globally for tin as:

$$H_2 + Sn[O_x]_{Sn} \longleftrightarrow H_2O \tag{4.2}$$

For a tin anode, Reaction 4.2 is believed to be confined to a thin region near the interface given that available data shows hydrogen being slightly soluble in liquid tin[8]. Should Reaction 4.2 be sluggish relative to oxygen transport in the liquid metal, a build-up of $Sn[O_x]_{Sn}$ will occur relative to equilibrium conditions. Using Equ.s 2.3b and 2.4, the effect of $Sn[O_x]_{Sn}$ on cell voltage can be written as:

$$E_N = \frac{RT}{4F} \ln(\frac{a_{O2-C}}{a_{O2-M}}) \tag{4.3}$$

For a finite change in the activity of a_{O2-M} in the liquid anode (say above the equilibrium concentration), the change in cell voltage will be:

$$\Delta E_N = \eta_I = \frac{RT}{4F} \ln(\frac{a_{O2-M}}{a_{O2-M}^*}) \tag{4.4}$$

where η_I is the interfacial loss, and a_{O2-M}^* is the activity of dissolved oxygen at equilibrium (zero current load). Hence, under current loading, the activity of dissolved oxygen at the interface will increase above equilibrium conditions thereby causing a reduction in cell voltage. If we assume a linear variation of dissolved oxygen activity with current load, e.g., $a_{O2-M} = a_{O2-M}^* (1 + i/i_{lx})$, then Equ. 4.4 becomes:

$$\eta_I = \frac{RT}{2F} \ln(1 + \frac{i}{i_{lx}}) \tag{4.5}$$

In Equ. 4.5, i_{Ix} can be interpreted to be the effective exchange current density for the interface kinetics. As long as the current load, i, is much lower than this interface exchange rate current, the loss due to this mechanism will be small. It remains to be seen whether a linear relation for the variation of activity of oxygen at the interface is valid, but given the first order relation shown in Equ. 4.2 for dissolved oxygen, it is considered a viable first-order approximation.

Finally, diffusion transport losses are likely to be different for the liquid anode-SOFC system over the conventional porous cermet technology. For example, in the tin-SOFC system, oxygen is transported via $[O]_d$ (dissolved oxygen) and SnO_x (metal oxide) depending on the amount of oxygen in the tin (see phase diagram in Figure 4). These species will migrate through the thickness of the tin anode from the high oxygen levels near the electrolyte, out to the low oxygen levels near the gas interface. Over the path of diffusion, any SnO_x may convert to $Sn + x[O]_d$ should the local activity of oxygen decrease to below a mole fraction of 0.02. It is presently unclear how exactly this multiphase transport of oxygen will affect the mathematical model for the anode concentration overpotential. It is clear, however, that whereas conventional SOFC fuel cell models could assume constant diffusion coefficients (over at least typical ranges of current density), the liquid anode-SOFC fuel cell will likely have complex diffusivity behavior, and will in general be current dependent.

5.0 EXPERIMENTAL DATA

The objective of the experimental research being performed at NETL is to understand the fundamental behavior of an SOFC operating with a liquid metal anode. Two goals for this research are: (1) to clarify the oxygen transport mechanism through the liquid tin; and (2) to separate and quantify the gas-anode and anode-electrolyte interfacial resistance. Identifying the rate-limiting step within the anode will guide the design of a high performance liquid tin anode SOFC.

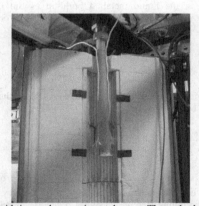

Figure 6. Liquid tin anode experimental setup. The cathode is exposed to open air and the liquid tin inside the zirconia tube is kept under a hydrogen flow or is gravity-fed solid fuel through a hopper (not shown) at the top of the tube.

As the current goal is to measure kinetic parameters of the anode and not to maximum performance per se, the sample cell was designed with a simple geometry that simplifies modeling and allows for easy changes in electrode dimensions, fuel flow, etc. Our experimental setup utilizes a 1.25 inch diameter electrolyte-supported button cell with a screen-printed LSM cathode and a thick molten layer of tin contained within a zirconia tube, as shown in Figure 6. Both YSZ and HionicTM (a scandia-stabilized zirconia purchased from NexTech Materials, Ltd.) electrolytes of ~120 μm thickness are being tested. The electrolyte support is attached to the end of a zirconia tube using Aremco 885 ceramic sealant. Current collection on the screen-printed cathode is performed using a platinum mesh attached with a platinum ink with silver lead wires. For current collection on the anode side, a graphite rod is submerged in the 6 mm thick 18 mm diameter molten tin. The free end of the graphite rod is wrapped in two silver lead wires—one for delivering current and the other for measuring cell potential. An alumina tube inside the larger zirconia tube acts as the anode inlet line, delivering fuel to within about 8 mm from the top of the liquid metal-gas interface, with a fuel feed of either hydrogen gas or activated carbon. The latter is delivered through a lockhopper attached to the top of the tube. The cell potential was continuously monitored using an Amrel ZVL electronic load. Electrochemical performance data were collected with a Solartron 1260/1287 Frequency Response Analyzer/Electrochemical Interface using a two-electrode configuration.

Figures 7 and 8 shows typical electrochemical impedance spectra from a liquid tin anode SOFC with a Hionic™ electrolyte support. For Figure 7, under 400 sccm hydrogen fuel at 850°C and 900°C, a small semicircular loop can be discerned in the higher frequency (~2 kHz) range, followed by a long tail in the lower frequency range (~ 10 mHz). While the frequency window of the high frequency loop shifts only slightly with temperature, its diameter almost doubles from 0.08 Ω to 0.15 Ω. Further tests are currently being performed to identify whether this loop can be attributed to processes in the cathode or anode.

In Figure 8, the spectra of hydrogen fueled operation is compared to the spectra of carbon fueled operation, both at 850 °C. Here, the lower frequency behavior is believed to be related to anode interfacial polarization at the electrode-gas interface, as the magnitude of this loop increases by an order of magnitude when the fuel is switched from flowing hydrogen to activated carbon (Calgon). As experimental issues related to maintaining a good seal on the tube and keeping the electrolyte from cracking are overcome, the lifetime of the samples will increase long enough to perform all the atmosphere-, polarization-, and temperature-dependent measurements needed to better characterize the cell.

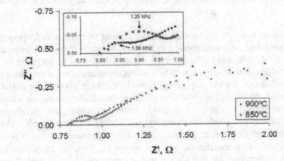

Figure 7. Impedance spectra from liquid tin/Hionic™/LSM SOFC under H_2 flow at 850° C and 900° C. The frequency labels in the inset indicate the frequency at the local maximum of the high frequency loop at both temperatures.

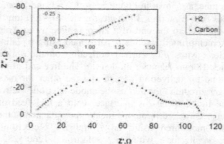

Figure 8. Impedance spectra from a liquid tin/Hionic™/LSM SOFC at 850°C under H_2 flow and when loaded with activated carbon.

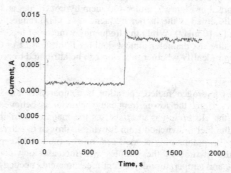

Figure 9. Current response to a 20 mV potential step from OCV of liquid tin anode SOFC under H_2 flow at 900° C.

To study oxygen transport through the molten tin, the current through a cell operated at 900 °C was monitored as a small potential step was applied to the cell. Figure 9 shows the average of 25 cycles of stepping to an OCV of 0.832 V and holding for 15 minutes followed by stepping to 20 mV off OCV (fuel cell mode) and holding for 15 minutes. While all mechanisms controlling the current have yet to be identified, it is possible that for the step away from OCV (at ca. 1000s), the initial spike and decay in the current corresponds to the initial oxidation of tin at the electrolyte/anode interface followed by the transport of the oxygen throughout the anode before reaching the steady-state condition as tin/oxygen is reduced at the gas/anode interface by fuel (either H_2 or carbon). Analysis of the data shows a transient characteristic time, τ, of about 240 s for this process. From a preliminary 1st order analysis of an *effective* diffusion coefficient, one has: $l^2/\tau = 1E-3 \ cm^2/s$. However, this value is about an order of magnitude higher than for that shown in the literature,[25],[26]. In part, this discrepancy may be due to the presence of dissolved hydrogen as well as induced convection within the tin from the gas supply, either of which might provide an effective diffusion length, *l*, much shorter than the thickness of the tin layer. To resolve the true oxygen diffusion effects, a detailed transport model may be required to account for the multiple reaction and transport mechanisms present. At present we are verifying that the observed behavior is indeed linked to oxygen transport through the cell by varying the electrode thickness, performing the measurement at multiple temperatures, and comparing the calculated oxygen diffusion coefficients and activation energies with those reported elsewhere[25].

It must further be noted that the OCVs obtained by our first samples were lower than that of both the oxidation of hydrogen and of tin. For example, for the data shown in Fig. 9, the OCV under hydrogen flow (with ~3% H_2O) at 900°C was 0.83 V. The hydrogen Nernst potential at 900°C is 1.08V, while that for the oxidation of tin to tin dioxide is 0.87 V. We believe the lower OCV for the sample is primarily the result of leaks in the sealant holding the cell to the sample tube. The observation of SnO_2 (confirmed by XRD) along the circumference of the tin anode where the sealant was applied supports this conclusion. We are currently examining different sealants (e.g., Aremco Ceramabond[TM] 552, Flexbar Autostic) and different cell configurations to overcome this problem and at present operate at an OCV slightly above the tin oxidation potential.

6.0 SUMMARY

Liquid tin anode (LTA-SOFC) technology is being studied by NETL for its ability to directly convert coal to electricity. Key to the success of the development of this technololgy is a detailed assessment for the electrochemical activity and oxygen diffusion within the liquid tin. In addition, the fundamental thermodynamic operation of such a concept needs to be properly analyzed. Initial research efforts to characterize the tin electrochemistry on a button cell at 900° C and with a tin thickness of 6mm show that both the temperature and the fuel type control the EIS signatures, and, hence, the various overpotentials present. Test procedures are being developed at NETL to also characterize the diffusion behavior of the oxygen in the tin. Preliminary 1st order analysis of the data provides evidence that a transient characteristic time of about 240s for reaching steady state following a voltage step change and for the hardware setup used. This corresponds to an *effective* oxygen diffusion coefficient of about 1E-3 cm^2/s which suggests that significant additional effects are present that modify a purely oxygen diffusion process.

Acknowledgements

The authors wish to thank Dr. Nguyen Minh for helpful comments made to support this work.

REFERENCES

[1] World Energy Outlook 2008, OECD/IEA, International Energy Agency, (2008).
[2] US DOE. Office of Fossil Energy. National Energy Technology Laboratory. Carbon Sequestration Technology Roadmap and Program Plan. (2007).
[3] J. Gale, S. Bachu, O. Bolland, Z. Xue, To Store or Not to Store?, *Int. J. Greenhouse Gas Control*, Volume 1, Issue 1, p. 1, April (2007).
[4] T. Tao, Introduction of Liquid Anode/Solid Oxide Electrolyte Fuel Cell and its Direct Energy Conversion Using Waste Plastics, Proceedings of the International Symposium 9th Solid Oxide Fuel Cell, Quebec, Canada, 353-362, May (2005).
[5] T. Tao, M. Slaney, L. Bateman, and J. Bently, Anode Polarization in Liquid Tin Anode Solid Oxide Fuel Cell, *ECS Trans.*, 7, (1) 1389 (2007).
[6] D.R. Gaskell, *Introduction to the Thermodynamics of Materials*, Taylor and Francis Books, Inc., p. 541 (2003).
[7] D.R. Gaskell, *Introduction to the Thermodynamics of Materials*, Taylor and Francis Books, Inc., p. 268, (2003).
[8] *Chemistry of Tin*, ed. by P.J. Smith, Blackie Academic & Professional, p. 6, (1998).
[9] *Handbook of Chemistry and Physics*, CRC Press, INC., ed. 67, (1986).

[10] S. Cahen, N. David, J.M. Fiorani, A. Maˆıtre, M. Vilasi, Thermodynamic modelling of the O–Sn system, *Thermochimica Acta* 403, 275–285, (2003).

[11] P.A. Wright, *Extractive Metallurgy of Tin*, Amsterdam, London, New York, Elsevier Publishing Company, (1966).

[12] *The Properties of Tin*, Tin Research Institute, International Tin Research and Development Council (1954).

[13] E. Ricci, L. Nanni, A. Passerone, Oxygen transport and dynamic surface tension of liquid metals, *Phil. Trans. R. Soc. Lond. A* 356, p. 857-70, (1998).

[14] *Constitution of Binary Alloys*, 2nd ed., McGraw-Hill, p1066, (1985).

[15] T.N. Belford and C. B. Alcock, *Trans. Faraday Soc.*, 61, 443-453, (1965).

[16] E.H.P. Cordfunke and R.J.M. Konings, Thermochem. Data for Reactor Materials and Fission Products, North-Holland, Amsterdam, (1990).

[17] I. Barin, O. Knacke, and O. Kubaschewski, Thermochemical Properties of Inorganic Substances, Springer-Verlag, Berlin, (1977).

[18] Adapted from T.J. Heames, D.A. Williams, N.E. Bixier, A.J. Grimley, C.J. Wheatley, N.A. Johns, P. Domagala, L.W. Dickson, C.A. Alexander, L. Osborne-Lee, S. Zawadzki, J. Rest, A.Mason and R.Y. Lee, Victoria: A Mechanistic Model of Radionuclide Behaviour in the Reactor Coolant System Under Severe Accident Conditions, Sandia National Laboratories (NUREG/CR-5545, SADN90-0756, Rev.1), (Dec. 1992).

[19] *Selected Values of Chemical Thermodynamic Properties*, National Bureau of Standards Series 270, D.D. Wagman et al., U.S. Department of Commerce, Washington, 1968-1971. Appended to: "Thermochemical Properties of Inorganic Substances", I. Barin, O. Knacke, and O. Kubaschewski, Springer-Verlag, Berlin, (1977).

[20] I. Barin, Thermochemical Data of Pure Substances, VCH, Weinheim, Germany (1989).

[21] M.G. Bonicelli, G. Ceccaroni, F. Gauzzi, and G. Mariano, Solidification of Metallic Tin Dispersed in Phase, *Thermochemica Acta*, 430, p. 95-99, (2005).

[22] Z.M. Zarzebeski, and J. P. Marton, Physical Properties of SnO_2 Materials. II. Electrical Properties, *J. Electrochem. Soc.*, 123 [9] 299c-309c (1976).

[23] B. Kamp, R. Merkle, J. Maier, Chemical Diffusion of Oxygen in Tin Dioxide, *Sensors and Actuators B*, 77, p. 534-42, (2001).

[24] I.M.B. Nielsen, C.L. Janssen, M.D. Allendorf, Ab initio Predictions for Thermochemical Parameters for Tin-oxygen Compounds, *J. Physical Chemistry A*, 107, p. 5122-27, (2003).

[25] H. Chou, T.C. Chow, S.F. Tsay, H.S. Chen, Diffusivity of Oxygen in Liquid Sn and $Ba_{0.35}Cu_{0.65}$ Alloys, *J. Electrochem. Soc.*, 142(6) pp.1814-1819, (1995).

[26] T. Itami, T. Masaki, H. Aoki, S. Munerjiri, M. Uchida, S. Matsumoto, K. Kamiyama, K. Hoshino, Self-diffusion Under Microgravitiy and Structure of Group IVB liquids, *J. Non-Crystalline Solids*, 312-314, p. 177-81, (2002).

A NO CHAMBER FUEL CELL USING ETHANOL AS FLAME

Kang Wang[1], Jeongmin Ahn[1*], Zongping Shao[2]

[1] Mechanical Engineering, Washington State University, Pullman, WA 99164, United States
[2] State Key Laboratory of Materials-Oriented Chemical Engineering, Nanjing University of Technology, No.5 Xin Mofan Road., Nanjing 210009, P.R. China

ABSTRACT:
A no-chamber solid-oxide fuel cell operated on a fuel-rich ethanol flame was reported. Heat produced from the combustion of ethanol thermally sustained the fuel cell at a temperature of 500-830 °C. Considerable amounts of hydrogen and carbon monoxide were also produced during the fuel-rich combustion which provided the direct fuels for the fuel cell. The location of the fuel cell with respect to the flame was found to have a significant effect on the fuel cell temperature and performance. The highest power density was achieved when the anode was exposed to the inner flame. By modifying the Ni+$Sm_{0.2}Ce_{0.8}O_{1.9}$ (SDC) anode with a thin Ru/SDC catalytic layer, the fuel cell envisaged not only an increase of the peak power density to ~ 200 mW cm^{-2} but also a significant improvement of the anodic coking resistance.

1. INTRODUCTION

Fuel cells represent one of the cleanest, most efficient and versatile technologies for chemical-to-electrical energy conversion [1]. Among many fuel cell types, solid-oxide fuel cell (SOFC) as a high-temperature electricity-generating device has received considerable attention due to its high efficiency and fuel flexibility [2, 3].

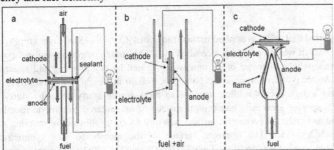

Fig. 1 Schematic of fuel cells: (a) dual-chamber SOFC; (b) single-chamber SOFC; (c) flame fuel cell

Conventional SOFCs are operated in a dual-chamber configuration (**Fig. 1a**), in which the cell is separated into two compartments with the help of sealant and interconnect with the anode chamber supplied with fuel and cathode chamber with air. This configuration is the most common for

* Corresponding author, Tel: 509-335-7711
  E-mail: ahn@mme.wsu.edu

large-scale applications of SOFC, which has the minimum requirement for the catalytic selectivity of electrodes. However the thermal expansion mismatch between cell components and sealant may introduce large internal stress during the heating and cooling processes. Therefore, normally the dual-chamber configuration is not considered for portable application in which frequent and rapid start-up and shut-down are necessary.

More recently, the concept of single-chamber solid-oxide fuel cell (SC-SOFC) was proposed (**Fig. 1b**) [4-10]. It is in a sealant-free configuration with both electrodes exposed to the same premixed fuel-air mixture. The performance of the SC-SOFC is based on the different catalytic selectivity of anode and cathode towards the fuel-oxidant mixture. Therefore, the anode and cathode materials should be carefully designed. Non-ideal behavior of electrodes could decrease the fuel cell efficiency significantly.

Very recently, the innovative concept of a direct-flame fuel cell based on SOFC, operated in a no-chamber mode, has also been proposed (**Fig. 1c**) [11,12]. As compared to SC-SOFC, the flame fuel cell configuration is even more simplified. The fuel-rich flame provides not only the fuels but also the heat for thermally sustaining the fuel cell. No external thermal management is then necessary. The advantages of the direct-flame fuel cell include ultra-simple cell configuration and highly-flexible fuel selection since the intermediate flame components are similar for all kinds of hydrocarbons. Furthermore, similar to SC-SOFC, it is capable for rapid start-up and ideal for portable applications. However, the performance of the direct-flame fuel cell is still poor up to now [11,12] which hinders its practical applications.

On the other hand, ethanol is an ideal fuel, which is a nontoxic, renewable and easily reachable resource with high energy density and in liquid state at normal conditions. The application of ethanol as a fuel for fuel cell has attracted increasing attention recently [13-18]. The direct ethanol fuel cell based on low-temperature polymer electrolyte membrane fuel cell (PEMFC) has been exploited in the past years. Unfortunately, it typically delivered poor performance [13-16]. In this paper, we report a coking-free flame fuel cell with improved anode by applying ethanol as fuel that doubled the performance as compared with ethanol fuel cell based on PEMFC. It may find potential applications in portable power generation.

2. EXPERIMENTAL

An electrolyte-supported fuel cell was applied in this study. $Sm_{0.2}Ce_{0.8}O_{1.9}$ (SDC) was employed as the electrolyte material, which was prepared by a combined EDTA-citrate complexing sol-gel process [19]. Disk-shape membrane with a diameter of 15 mm was prepared by dry pressing and sintered at 1350 °C for 5 h in air. Anode powder composed of NiO+SDC (60:40 in weight) was first mixed with ethanol and glycol using a high energy milling machine to form a even slurry, then the slurry was spray-deposited (using a sprying gun) onto one central surface of the sintered electrolyte membrane with a rectangular shape and effective geometric surface area of 0.48 cm^2, followed by sintering at the same temperature for 5 h in air. The opposite surface of the membrane was symmetrically spray-deposited with a thin layer of $Ba_{0.5}Sr_{0.5}Co_{0.8}Fe_{0.2}O_{3-\delta}$ (BSCF) + SDC (70:30 in weight) composite and sintered at 1000 °C for 5 h in air. For some fuel cells, the anode surface was further screen-printed with a layer of Ru/SDC (8 wt.%) catalyst and fired at 800 °C in air.

Table 1
Detailed compositions, processing parameters and selected properties of the fuel cells

Componens	Materials	Fabrication methods	Sintered temp. (°C)	Sintered time (h)	Thickness (μm)
Electrolyte	SDC	dry pressing	1350	5	400
Anode	SDC+NiO	spraying	1350	5	50
Cathode	BSCF+SDC	spraying	1000	5	15
Catalytic layer	Ru/SDC	Screen printing	850	0.5	40

Table 1 shows the detailed fuel cell compositions and processing parameters. The surface morphologies of the fuel cells were observed by a scanning electron microscope (SEM, Quanra-2000). Shown in **Fig. 2** are the typical cross-sectional morphologies of the cell.

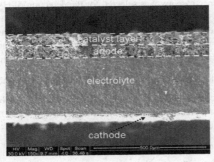

Fig. 2 The typical cross sectional morphologies of the electrolyte-supported fuel cell with a Ru+SDC catalytic layer

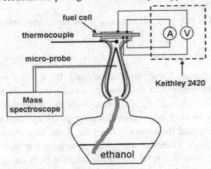

Fig. 3 Experimental setup for a direct-ethanol flame fuel cell

A normal ethanol lamp was applied to provide the flame for the fuel cells. **Fig. 3** shows a schematic of the fuel cell reactor. Here, 99.97% ethanol was applied as the fuel with a density of 0.79 g cm^{-3} (20 °C). The burning rate of the ethanol flame was kept constant at 0.19 g min^{-1}. The fuel cell was located on top of the flame with the anode facing the flame front and the cathode breathing ambient air. Silver paste and silver wires were applied as the current collector to minimize possible catalytic activity for ethanol reforming. A K-type thermocouple, penetrating into the flame or attached to the fuel cell surface, was applied for measuring the flame or fuel cell temperature. The fuel cell performance was tested by *I-V* characterization using a digital sourcemeter (Keithley 2420) interfaced with computer for data acquisition.

The flame gases were sampled to a mass spectroscope (Hiden QIC 20) via a micro-tube for in-situ composition analysis. The sampling rate was so small that it had negligible effect on the flame shape and composition. The mass spectroscope was pumped to a pressure of $\approx 1 \times 10^{-5}$ Torr using a turbomolecular pump. Note that water and ethanol were removed from the products before they reached the mass spectrometer in order to avoid poisoning. The micro-tube first sampled the ambient air for about 400 seconds before it quickly penetrated into the flame for flame composition analysis.

The peak at m/z of 2 in mass spectroscopy was selected for the detection of H_2. Although water and ethanol could also produce H_2^+ fragments, their contributions were negligible since they were pre-removed before sampling into the mass spectroscope. The peak at m/z of 44 was selected for the detection of CO_2. As to CO, since N_2 and CO have the same molecule weight of 28 and CO_2 could also produce CO^+ fragment, we then can not simply take the peak at m/z of 28 for CO. In order to get the CO content from the peak at m/z of 28 (C_{det}), the contributions from N_2 and CO_2 should be deducted. We know that the peak at m/z of 14 contributed from the characteristic fragment of N_2, *i.e.*, N^+, which has a fixed relative intensity of 7.2% to that of N_2^+ at m/z of 28 for a pure N_2 gas. On the other hand, the peak intensity of CO^+ from the cleaving of CO_2 is 11.4% that of CO_2^+ at m/z of 44 for a pure CO_2 gas. The CO content (C_{CO}) was then calculated as follows:

$$C_{CO} = C_{det} - C_{N2} - 0.114 * C_{CO2} \qquad (1)$$
$$C_{N2} = 100 * C_N/7.2 \qquad (2)$$

where C_N is the detection content of N at m/z of 14, and C_{CO2} is the detection content of CO_2 at m/z of 44, in mass spectroscopy, respectively.

3. RESULTS AND DISCUSSION

As shown in **Fig. 4a**, the ethanol flame is characterized by a three-layer structure with distinguished layer colors. Across the flame from the center to outside, the layers are named as central flame, inner flame and outer flame, respectively. Different from the premixed fuel-air flame, the ethanol flame investigated here was created by an inter-diffusion of ethanol and oxygen. During the combustion, ethanol continuously diffused from the center of the flame to ambient atmosphere while oxygen diffused in the reverse direction. Therefore a gradient of oxygen and fuel concentrations was built across the flame with the oxygen-to-fuel ratio increasing steadily from the center to the outside. Positioning the fuel cell on top of the flame resulted in a change of the flame shape as shown in **Fig. 4b**.

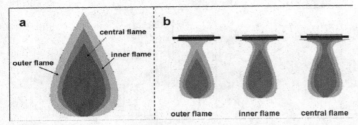

Fig. 4 Schematic of the shape of an ethanol flame, (a) without fuel cell and (b) with fuel cell over the flame at different positions

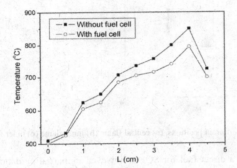

Fig. 5 Location-related temperature profiles of flame with respect to flame with and without the fuel cell disk over the top of the flame

A thermocouple was applied to probe the temperature of the flame at different positions with and without the fuel cell disk on top of the flame. As shown in **Fig. 5**, the flame temperature reached 500-830 °C, depending on the position in the flame. An increase in the temperature was observed across the flame from the center to the outside. Positioning the fuel cell on top of the flame resulted in a decrease of the flame temperature of about 10-30 °C. Such a decrease is easy to understand since the fuel cell increased the heat loss from surface radiation and also presents a physical barrier for oxygen diffusion. Interestingly, such a fuel cell temperature range was just ideal for operating the doped-ceria based electrolyte. Long-term stability test demonstrated that the cell temperature was stable if the ethanol combustion rate kept constant and there was no environmental disturbance, which suggests that the ethanol flame can be applied efficiently as a heater for thermally sustaining the fuel cell, which then omits the requirement for external thermal management and greatly simplifies the fuel cell system.

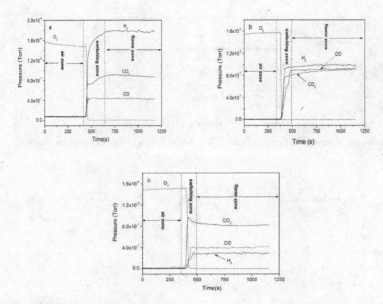

Fig. 6 Gas compositions of the flame at different positions, (a) central flame (b) inner flame (c) outer flame

Ethanol by itself can serve as a direct fuel for SOFC; however, its partial oxidation products of syngas ($CO+H_2$) are by far much more active electrochemically. Another advantage of syngas over ethanol as fuel is that the fuel cell anode is expected to be much less prone to carbon coking. Therefore ethanol reforming products are preferred as the fuels for SOFC. It is well known that the flame can act as a fuel reformer. The approach of using fuel-rich flames for the production of synthesis gas via the partial oxidation has been demonstrated by several authors, in particular with the help of porous combustors [20-22]. It was reported that the reaction products of fuel-rich pre-mixed hydrocarbon flame consist of a mixture of mainly N_2, CO_2, H_2O, CO and H_2 [23]. This is believed to be true for all kind of hydrocarbons including gaseous, liquid and solid fuels [12]. Current ethanol flame, especially for the inner and central flames, could be treated as a fuel-rich flame with steadily increasing fuel-to-oxygen ratio across the flame from outside to center.

In order to assess the nature of the fuel species available for the SOFC of the ethanol flame, the compositions of the ethanol flame were experimentally analyzed by a micro-quartz tube probe in connection with a mass spectroscope. **Fig. 6** shows the gas compositions of the flame at positions about 1cm (central flame), 2.4 cm (inner flame) and 4.0 cm (outer flame) away from the center, respectively. "Air zone" and "flame zone" mean the zones at which mass spectroscope was pumping the air and the flame gases, respectively. The switching zone means the zone of the mass spectroscope transitioning from probing air to probing flame products. It clearly demonstrated there were considerable amount of H_2 and CO produced within the flame at steady state. Therefore, the fuel-rich ethanol combustion flame acted efficiently as a fuel reformer in this study. The deeper into of the flame, the higher the H_2 concentration and the lower the CO_2 concentration were observed. Such phenomena can be explained by the increased fuel-to-oxygen ratio across the flame.

During the ethanol combustion process many reactions are possible including:

$$CH_3CH_2OH+3O_2 \rightarrow 3H_2O+2CO_2 \qquad\qquad \Delta H^\theta_{298}=-1365.5 \text{ kJ mol}^{-1} \qquad (3)$$

$$CH_3CH_2OH +H_2O \rightarrow 4H_2+2CO \qquad\qquad \Delta H^\theta_{298}=256.8 \text{ kJ mol}^{-1} \qquad (4)$$

$$CH_3CH_2OH +1/2O_2 \rightarrow 3H_2+2CO \qquad\qquad \Delta H^\theta_{298}=14.1 \text{ kJ mol}^{-1} \qquad (5)$$

$$CH_3CH_2OH +2H_2O+1/2O_2 \rightarrow 5H_2+2CO_2 \qquad\qquad \Delta H^\theta_{298}=-68.5 \text{ kJ mol}^{-1}. \qquad (6)$$

Deep oxidation of ethanol would produce a huge amount of heat (eq.3), which explained the highest temperature for the outer flame where the oxygen concentration was the highest due to the easiest diffusion of oxygen from the ambient air. Under the fuel-rich condition, the combustion would prefer reactions (4), (5) and (6). Consequently much less heat while more hydrogen was produced, and therefore a decrease in flame temperature was expected.

Once the SOFC was positioned over the ethanol flame with the anode facing the flame front, a positive open circuit voltage (OCV) was established in seconds while the fuel cell performance reached steady state in minutes. **Fig. 7** shows the *I-V* curves of the fuel cells (stable performance) without and with the catalytic layer on top of the anode. For both cases, the OCVs experienced an increase with the fuel cell moving downward from the outer flame to the central flame. An OCV of 0.75, 0.70 and 0.65 V was observed (**Fig. 7(a)**), for the fuel cell located over the central flame, the inner flame and the outer flame, respectively. Based on the results in **Fig. 4 & 5**, the increase in OCV should be mainly attributed to the decrease in fuel cell temperature and the increase in fuel concentration over the anode side. At higher temperatures, the electronic conductivity of the SDC electrolyte became more significant, which resulted in the internal electronic shorting of the doped ceria electrolyte and consequently in a decreasing OCV. The maximum OCV (~0.75 V) observed in the current flame fuel cell was slightly higher than that obtainable in an SCFC employing the same fuel cell components [5, 6]. In the flame fuel cell, the cathode actually breathes the air, which ensures higher oxygen partial pressure at the cathode than that in the SCFC. In this manner, the flame fuel cell might be treated like a kind of internal reforming dual-chamber SOFC.

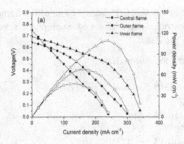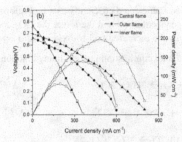

Fig. 7 The dependence of cell voltage and power density on current density of the fuel cells, (a) without and (b) with the catalytic layer on the top of the anode

A peak power density of 45, 105 and 63 mW/cm² was obtained for the fuel cell without the catalytic layer and with the anode facing the central, inner and outer flames, respectively. They are comparable to the results reported by other groups with methane-based flame-powered fuel cells[11, 12]. Although the central flame had the highest concentration of hydrogen as shown in **Fig.6**, it also had the lowest fuel cell temperature. The low cell temperature significantly reduced the electrode activity and increased the ohmic resistance of the electrolyte. On the other hand, the outer flame had the highest fuel cell temperature, but the hydrogen concentration was low due to the presence of a large amount of oxygen. Comparatively, the inner flame demonstrated the highest fuel cell performance since it had both an appropriate temperature and a favorable gas composition.

The OCV was only slightly improved when the anode surface was deposited with a thin catalytic layer. A value of 0.66, 0.71, and 0.77 V was obtained for the cases of outer, inner and central flames, respectively (**Fig.7(b)**), which was only 0.02, 0.01 and 0.01V higher than the case without the catalytic layer. However, the catalytic layer greatly enhanced the fuel cell performance, evidenced by a peak power density of 60, 200 and 130 mW/cm² corresponding to the central, inner and outer flames, respectively. The significant improvement in performance could be due to the catalytic activity of the Ru/SDC catalytic layer towards ethanol reformation and partial oxidation. It is well known that the Ru/CeO₂-based catalysts have superior catalytic activities for hydrocarbon partial oxidation and reforming[25-27]. The deposition of the Ru/CeO₂-based catalyst over the anode might then promote the reactions (3)-(6). Kinetically, there should be a considerable amount of unreacted ethanol reaching the anode during the fuel cell operation after combustion. The Ru/SDC catalytic layer heterogeneously catalyzed the ethanol to syngas (CO+H₂) over the anode surface by reforming ethanol with the CO₂ and H₂O formed in-situ or by partially oxidizing ethanol with the oxygen diffusing in from the ambient atmosphere. Due to the enhanced production of syngas, which is much more active electrochemically than ethanol, an increase in fuel cell performance should be expected.

By comparison with the SOFC in this study, the direct ethanol fuel cell based on polymer electrolyte membrane has also received increased attention due to the nontoxic and renewable characteristics of ethanol. However, up to now, its peak power density has been typically lower than 100 mW/cm² [13-16]. Our data shows that the direct flame SOFC in this study delivers a much better performance.

Fig. 8 SEM surface morphologies of the anode surface, (a) without and (b) with the Ru+SDC catalytic layer after test

Despite the remarkable improvement in power density, coking remains a problem for the flame fuel cell in some cases. After an operation for more than 10 hours, the anode was examined by SEM. **Fig.8** shows the SEM morphologies of the anode surface without and with the catalytic layer. Serious coking was observed over the fuel cell anode with bare Ni+SDC, which could be explained by the well-known fact that nickel catalyst promotes hydrocarbon cracking [28]. When Ru/SDC catalytic layer was applied however, the anode demonstrated no carbon deposition after the fuel cell test. This indicates that Ru/SDC greatly improved the coking resistance of the anode, and similar super coking resistant properties of the Ru/CeO$_2$-based catalytic layer have also been reported in literature [29-31]. So we believe that this improved carbon-coking resistance of the anode is closely related to the capability of the ruthenium catalyst to promote the elimination of the carbon coking by the following reaction

$$C+H_2O \rightarrow CO+H_2 \qquad (7)$$

4. CONCLUSIONS

The ethanol flame, which serves as a fuel reformer while at the same time providing the required heat for the fuel cell operation, was found to be an ideal power source for the flame fuel cells based on the SDC electrolyte. It was found that the fuel cell temperature and performance strongly depended on the location of the fuel cell with respect to the flame, and the performance decreases in the order of inner, outer and central flames for fuel cells both with and without the catalytic layer. A maximum power density of 200 mW/cm^2 was achieved with the catalytic-layer-deposited anode exposed to the inner flame. The functional catalytic layer demonstrated beneficial effects on improving both the performance and the coking resistance of the fuel cell. Considering the high performance and the ultra-simple configuration, the direct ethanol flame fuel cell reported here may have the potential for applications in portable power generation in future.

However, a few challenges remain for this type of fuel cells and need to be addressed by future research: (1) Thermal management. The uneven heating of the flame will induce the thermal stress within the fuel cell and may cause it to crack. (2) Fuel utilization. Most of the fuel is consumed by combustion and therefore the total electrical efficiency is low. (3) System design. Flames in real situations are generally hard to control and may often be unstable. Despite these challenges, flame fuel cells could find their potential applications in scenarios where the need for portable power generation overrides the requirement for high fuel utilization, and / or where heat and electricity co-generation is desired.

REFERENCES

[1]   B.C.H. Steele, A. Heinzel, Nature, 414 (2001) 345-352.

[2]   S.W. Tao, J.T.S. Irvine, Nat. Mater., 2 (2003) 320-323.

[3]   S. Park, J.M. Vohs, R.J. Gorte, Nature, 404 (2000) 265-267.

[4]   T. Hibino, A. Hashimoto, Science, 288 (2000) 2031-2033.

[5]   Z.P. Shao, C. Kwak, S.M. Haile, Solid State Ionics, 175 (2004) 39-46.

[6]   Z.P. Shao, S.M. Haile, Nature, 431 (2004) 170-173.

[7]   T. Suzuki, P. Jasinski, V. Petrovsky, H.U. Anderson, F. Dogan, J. Electrochem. Soc., 151 (2004) 1473-1476.

[8]   I.C. Stefan, C.P. Jacobson, S.J. Visco, L.C. De Jonghe, Electrochem. Solid-State Lett., 7 (2003) 198-200.

[9]   K. Wang, Z.P. Shao, Prog. Chem. (in Chinese), 19 (2007) 267-275.

[10]   Z.P. Shao, J. Mederos, W.C. Chueh, S.M. Haile, J. Power sources, 162 (2006) 589-596.

[11]   M. Horiuchi, S. Suganuma, M. Watanabe, J. Electrochem. Soc., 151 (2004) 1402-1405.

[12]   H. Kronemayer, D. Barzan, M. Horiuchi, S. Suganuma, Y. Tokutake, C. Schulz, W.G. Bessler, J. Power Sources, 166 (2007) 120-126.

[13]   G. Andreadis, P. Tsiakaras, Chem. Eng. Sci., 61 (2006) 7497-7508.

[14]   E. Peled, T. Duvdevani, A. Aharon, A. Melman, Electrochem. Solid State Lett., 4 (2001) 38-41.

[15]   C. Lamy, S. Rousseau, E.M. Belgsir, C. Coutanceau, J.-M. Léger, Electrochim. Acta, 49 (2004) 3901-3908.

[16]   J.R. Varcoe, R.C.T. Slade, E.L.H. Yee, S.D. Poynton, D.J. Driscoll, J. Power Sources, (2007), doi:10.1016/j.jpowsour.2007.04.068

[17]   B. Huang, S.R. Wang, R.Z. Liu, X.F. Ye, H.W. Nie, X.F. Sun, T.L. Wen, J. Power Sources, 167 (2007) 39-46.

[18]   S.L. Douvartzidesa, F.A. Coutelierisa, A.K. Deminb, P.E. Tsiakarasa, Int. J. Hydrogen Energy, 29 (2004) 375-379.

[19]   W. Zhou, Z.P. Shao, W.Q. Jin, J. Alloys Compds., 426 (2006) 368-374.

[20]   Y.Y. Qian, J.B. Chen, Z.Z. Wu, R.Y. Wang, J.H. Xu, Chinese J. Chem. Educ., 24 (2003) 39-41.

[21]   F.J. Weinberg, T.G. Bartleet, F.B. Carleton, P. Rimbotti, Combust. Flame, 72 (1988) 235-239.

[22]   D. Trimis, F. Durst, Combust. Sci. Technol., 121 (1996) 153-168.

[23]   W.M. Mathis Jr., J.L. Ellzey, Combust. Sci. Technol., 175 (2003) 825-839.

[24]   H. Pedersen-Mjaanes, L. Chan, E. Mastorakos , Int. J. Hydrogen Energy, 30 (2005) 579-592.

[25]   S. Hosokawa , S. Nogawa, M. Taniguchi, K. Utani, H. Kanai, S. Imamura, Appl. Cata. A, 288 (2005) 67-73.

[26]   L. Oliviero, J. Barbier Jr., D. Duprez, H. Wahyu, J.W. Ponton, I.S. Metcalfe, D. Mantzavinos, Appl. Catal. B, 35 (2001) 1-12.

[27]   Z.P. Shao, S.M. Haile, J. Ahn, P.D. Ronney, Z. Zhan, S.A. Barnett, Nature, 435 (2005) 795-798.

[28]   B.C.H. Steele, Solid State Ionics, 86–88 (1996) 1223-234.

[29]   K. Wang, R. Ran, Z.P. Shao, J. Power Sources, 170 (2007) 251-258.

[30]   Z.L. Zhan, S.A. Barnett, Science, 308 (2005) 844-847.

[31]   T. Hibino, A. Hashimoto, M. Yano, M. Suzuki, M. Sano, Electrochim. Acta, 48 (2003) 2531-2537.

Characterization/Testing

SURFACE ENHANCED RAMAN SPECTROSCOPY FOR NVESTIGATION OF SOFC CATHODES

Kevin S. Blinn, Harry W. Abernathy, and Meilin Liu
Georgia Institute of Technology, Department of Materials Science and Engineering
Atlanta, GA 30332

ABSTRACT

A profound understanding of SOFC cathode surface structure and chemistry under operating conditions is vital to unraveling the mechanisms of oxygen reduction, a critical step toward rational design of more efficient cathode materials. In this presentation, we report our recent findings in investigation into the surfaces of various cathode materials (e.g., LSM, LSC, SSC, and LSCF) using surface enhanced Raman spectroscopy (SERS), which can be much more sensitive than the conventional Raman methods. Cathode materials were made SERS-active by depositing drops of 20-nm Ag colloid on the material surface, through co-deposition of the materials with Ag nanoparticles by combustion chemical vapor deposition, or by short-term DC sputtering of Ag onto the cathode surface. Raman spectra collected from some of the SERS-active materials showed enhancement of more than one order of magnitude for peaks corresponding to lattice phonon modes and surface adsorbed oxygen species, demonstrating that SERS is a powerful tool for in situ characterization of surface structure and chemistry of electrode materials.

INTRODUCTION

Solid oxide fuel cells (SOFCs) hold great potential in the field of energy generation. Their high operating temperatures convey numerous advantages, such as low internal resistance and usable exhaust heat for driving secondary energy generation systems. SOFCs also allow for the use of cheaper catalysts than other types of fuel cells, and they are able to directly use hydrocarbon gases such as CO as fuel. With increasing amounts of research into SOFC technology, this alternative energy source grows ever closer to becoming a mainstream route for stationary power generation. One of the main technical challenges in the SOFC field is the reduction of the fuel cell's operating temperature. Reducing the SOFC operating temperature (to more intermediate values below 700°C) would impart advantages such as lower degradation of electrodes and decreased thermal stresses at component interfaces. Lowering the SOFC's operating temperature, however, decreases its conductivity and catalytic activity. The two factors that reduce conductivity in SOFCs are electrolyte bulk resistance and polarization resistance at the electrode-electrolyte interfaces. Electrolyte materials with higher conductivity coupled with thinner electrolyte layers have already been used to lower electrolyte bulk resistance in intermediate-temperature SOFCs[1-3]. The polarization resistance on the cathode side of the cell remains a chief hindrance to fuel cell performance. A deeper understanding of the oxygen reduction process, the key mechanism behind cathodic polarization resistance, would help overcome this barrier. Some consensus exists as to which types of events occur during the process. They include the adsorption of oxygen molecules, the dissociation of these molecules into atoms, the reduction of atoms and molecules to ions, and the incorporation of ions into the electrolyte[4]. The order and combination of the events that form a rate-limiting step for oxygen reduction is the subject of much debate, however[5]. Impedance spectroscopy is a highly useful tool for electrochemical analysis, and is therefore a ubiquitous methodology for oxygen reduction studies[5, 6]. Unfortunately, an impedance spectrum can quickly become convoluted by the influence of intricate charge and mass transfer processes, some of which the details may be unknown.

Characterization of cathode surface structures and chemistry would help uncover information missed by conventional electrochemical analysis. In-situ surface studies would provide even more information, as they would provide real-time insight as to what processes are actually occurring on the cathode during fuel cell operation. Unfortunately, the most widely used surface analysis methods based on electron spectroscopy such as X-ray photoelectron spectroscopy (XPS) and Auger electron spectroscopy (AES) are not suitable for such in-situ studies, as they require a certain degree of vacuum. SOFC cathodes would be exposed to near-atmospheric conditions in practical situations, so experiments involving samples exposed to similar oxygen partial pressures would be more ideal for investigating cathode surface processes.

One possible route for further analyzing SOFC cathode surface interactions is Raman spectroscopy, a proven versatile tool for characterizing materials in general. Raman spectroscopy can be used to identify different relevant molecular species that are present on a cathode's surface by their unique vibration modes, which can be observed through inelastic scattering of light. The use of Raman spectroscopy in the detection of species on oxide surfaces dates back to the 1970[7]. Of more particular interest is oxygen adsorption on catalysts, which has previously been investigated on CeO_2, CoO, and Pt[8-10]. Oxygen species interactions have even been demonstrated as observable by Raman spectroscopy even under high-temperature in-situ conditions. Itoh et al.[11] detected vibration modes at the TPB for a molten carbonate fuel cell (MCFC) presumably belonging to superoxide (O_2^-) and peroxide (O_2^{2-}) ions, which are key species in the oxygen reduction process. Furthermore, in-situ Raman spectroscopy has been used to characterize O_2- CeO_2 interactions[12], in which a CeO_2 surface was exposed to $^{16}O_2$ and $^{18}O_2$ isotope gases. Peaks in the 800-900 and 1000-1100 cm^{-1} were connected to peroxo- and superoxo-like species by coupling the spectroscopy results with DFT calculations. When similar techniques were initially employed for the sake of analyzing O_2 interactions with $La_{0.8}Sr_{0.2}O_{3-\delta}$ (LSM), a commonly studied SOFC cathode material, similar peaks were not discernible in the collected spectra. Thus, Raman signal enhancement may be necessary for the study of oxygen reduction on SOFC cathode materials.

The principal difficulty with using Raman spectroscopy for surface analysis is that in many cases, insufficient molecules of interest are available to impart enough Raman cross section to satisfy the sensitivity limits of Raman signal detection. While surfaces may have no shortage of molecules in general, some of these species may be of no interest to the investigator and would only serve to dilute the Raman spectra. Thus, rigorous cleaning of the surface of unwanted molecules might be required for successful analysis. Using this type of analysis for in-situ catalysis studies (i.e. SOFC cathode surface analysis) makes conditions even less ideal, as a high temperature environment does not compel species of interest that are weakly bonded to the surface to stay adsorbed for very long. Thus, enhancement of the Raman signal becomes compulsory for investigation of the oxygen reduction process on cathode material surfaces. In the 1970s, research groups investigating pyridine molecules adsorbed to roughened silver electrodes discovered that the silver enhanced the signal of the pyridine[13]. The concept behind this signal augmentation came to be known as surface-enhanced Raman scattering (SERS). The enhancement is thought to come mainly from a surface plasmon excitation phenomenon in which photons excite electron clouds in metallic structures 10-100 nm in size. These features create "hot spots" that enhance the signal of materials within several nanometers. SERS methods have lead to reports of signal enhancement factors of up to 14 orders of magnitude for single molecules[14]. From its inception until today, SERS has been used for a wide variety of studies and many methods for achieving the effect have been developed.

We do not know of any previously reported investigations involving the application of SERS for detecting species on SOFC cathode materials and enhancing material signals themselves that precede this one. Our own study utilizes metal colloids, co-deposited Ag nanoparticles, and discontinuous sputtered Ag films to produce SERS effects on popular SOFC cathode materials such as LSM, $La_{0.7}Sr_{0.3}Co_{0.2}Fe_{0.8}O_{3-\delta}$ (LSCF), $La_{0.6}Sr_{0.4}CoO_{3-\delta}$ (LSC), and $Sm_{0.5}Sr_{0.5}CoO_{3-\delta}$ (SSC). SERS application can result in the emergence of otherwise absent peaks and enhancements of existing smaller peaks of one or more orders of magnitude.

EXPERIMENTAL DETAILS

Sample Powder Preparation

LSM, LSCF, and SSC powders were either acquired commercially or synthesized by a solid state reaction (SSR) method. Precursors for SSR powder were ball-milled in ethanol with yttria-stabilized zirconia (YSZ) media for 24 hours. The mixed precursors were then calcined at 1050°C for 10 hours twice, with grinding and mixing following each firing. The resulting powders and commercially acquired powders were uniaxially pressed into cylindrical pellets 10 mm in diameter using 3 tons of pressure. Pressed pellets were fired at either 1200°C (SSC) or 1350°C (LSM and LSCF).

SERS Sample Preparation

In the first sample preparation method, 2-5 drops from a 20 nm Ag water-based colloid purchased from British Biocell International were applied to sintered SSC and LSCF sample pellets. The droplets were left to dry in air, after which the pellets were heat-treated at 325°C for 3 hours. The second method, combustion chemical vapor deposition (CCVD) was used to deposit Ag and LSC nanoparticles on a dense YSZ substrate simultaneously by spraying a solution of precursors through a flaming nozzle. CCVD methods are described in greater detail elsewhere[15]. The final method entailed the DC sputtering of a thin and discontinuous Ag films on the surfaces of polished LSM, LSCF, and SSC, using a 10-second sputtering time. The key for this route is that the sputtering time should be short enough to form small nanosized metal "islands" on the surface that are SERS active. Illustrations of these methods are displayed in Figure 1 below.

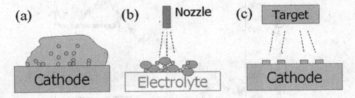

Figure 1. Deposition of Ag nanoparticles on cathodes using (a) colloids, (b) CCVD, and (c) DC sputtering.

Collection of Raman Spectra

Raman analysis was performed using a Renishaw RM-1000 Raman microscope at room temperature under atmospheric conditions. An excitation laser with a wavelength of 514 nm and an intensity of ~20 mW/μm^2 was used in all cases. In addition, spectra were collected from an Ag-sputter-treated LSCF pellet exposed to O_2 gas inside a special Raman sample chamber. The chamber was evacuated for 10 hours at 75°C, and then dry Ar gas was allowed to flow through the chamber at 400°C for 4 hours. Following this treatment, 20% O_2 gas, balanced by Ar, was released into the chamber at room temperature. As SERS "hot spots" are nanoscale features that cannot be readily observed by optical microscope, spectra were collected from rectangular grids of up to 2000 points with 2-3 μm between each point over a treated sample area using the mapping capabilities of the Raman microscope. Scans of 3-5 seconds were adequate for collecting spectra of acceptable signal-to-noise ratio when the SERS effect was observed.

RESULTS AND DISCUSSION

General Spectrum Collection

Most spectra collected consisted of a fluorescence signal from silver. However, spectra collected from 10% of the spots on the measured grid displayed signals attributable to SERS effects, on average. Examples of both types of spectra, which were taken from a colloid-treated LSCF pellet, are shown below in Figure 2. The relative scarcity of the SERS spots supports the theory that the effect only occurs in certain "hot spots" where the metal nanostructures happen to be in a favorable configuration. In the case of the colloid-treated samples, little to none of the spectra collected were characteristic of SERS spots when the sample was not heat-treated after colloid application and evaporation. Firing the treated samples may induce some flow of the silver nanoparticles, causing outcroppings or thin inter-particle connections to appear that induce SERS more effectively than before the heat treatment.

Spectra collected from the SERS hot spots contain several new peaks not seen in conventional spectra. Some of these peaks possibly originate from traces of surface contaminants. For example, the feature seen around 1350 cm^{-1} in the hot spot spectrum in Figure 2 potentially corresponds to disordered carbon[16]. The source of this contaminant carbon is possibly ethanol used to aid evaporation during the application of the colloids. The ethanol likely decomposed during firing. Different experiments and chemical calculations would be needed in order to rationally assign and truly verify all of the new peaks.

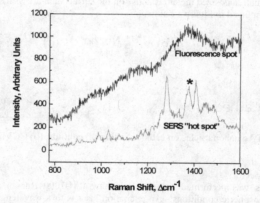

Figure 2. Characteristic Raman spectra collected from the area of LSCF pellet surface treated with 20 nm Ag colloid and heated to 325°C for 3 hours.

SERS Using CCVD with Ag

The peaks in the spectra collected from the LSC-Ag nanoparticle composites formed by CCVD were magnified by at least one order of magnitude when compared with those from the cathode materials alone, as shown in Figure 3. The four peaks indicated in spectra for LSCF alone are very weak, and perhaps arguably absent. They are quite prominent, however, for LSC combined with silver nanoparticles. Enhancement was only observed in the low wavenumber regime, however. Thus, these peaks are not likely surface species, but might be attributed to lattice modes within the LSC material itself.

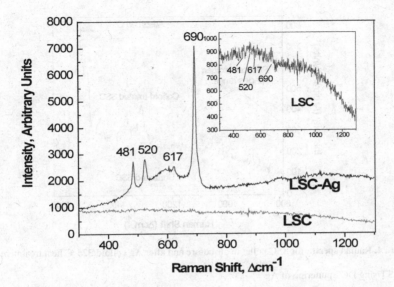

Figure 3. Raman spectra collected from LSC and CCVD-prepared LSC-Ag composite. The inset gives a closer look at the LSC spectrum.

SERS Using Ag Colloids

Displayed in Figure 4 below are Raman spectra collected from SSC samples in air at room temperature, before and after the Ag colloid treatment. 15-20 new peaks emerge in the SERS spectra. Some of the peaks in the 800-1100 cm^{-1} range potentially correspond to equilibrium reduced oxygen species adsorbed on the SSC surface, while the large peak near 1500 cm^{-1} can possibly be attributed to neutral surface oxygen molecules[8, 12]. The rest of the peaks may indicate other surface contaminants. Spectra like this one were only observed about 1-5% of the time for colloid samples, however.

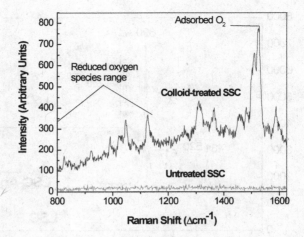

Figure 4. Raman spectra for SSC pellet in air before and after Ag colloid/325°C heat treatment.

SERS Using DC Sputtering of Ag

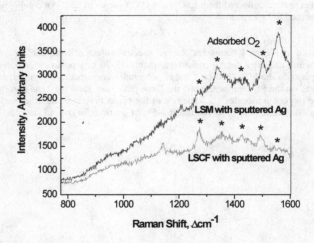

Figure 5. Raman spectra collected from LSM and LSCF pellets in air after DC sputtering of Ag.

Figure 5 above shows the results of Raman analysis performed on LSM and LSCF samples on which highly discontinuous Ag thin films were deposited by DC sputtering. Short-term sputtering could presumably deposit island-like features less than 100 nm in size in order to create SERS enhancement, although the size of the islands would require confirmation by microscopy with magnification greater than that of the Raman microscope. The spectra show indication of signal enhancement, although the potential reduced oxygen species features are not as apparent as with some of the spectra with signal enhanced by colloids. That being said, a larger percentage (~10%) of the sample points collected from the sputtered samples showed signs of SERS than those collected from colloidal samples. The enhanced adsorbed molecular oxygen peak was usually present with the sputtered samples, as well. As can be seen above, the same features in general can be found in spectra from LSCF and LSM samples, but with slight shifts and varying degrees of enhancement.

SERS Under Controlled Oxygen Atmosphere

Raman spectra collected following the evacuation, Ar gas exposure, heat treatments, and cooling to room temperature of an Ag-sputtered LSCF pellet in the special sample chamber did not seem to change in terms of general features from those collected from that type of sample beforehand. The fraction of SERS spectra observed per spectra collected, however, did significantly increase to about 15-20% after treatment. Again, the high temperature may have induced Ag flow that might be responsible for creating more nanoscale structures, producing a higher amount of SERS hot spots. The spectra obtained under various atmospheric conditions are shown in Figure 6 below. As can be observed in the figure, the peaks in the 800-1100 cm^{-1} became widely more pronounced upon oxygen exposure, supporting the claim that at least some of these features correspond to reduced oxygen species. In addition, the features marked with asterisks, which may correspond to disordered carbon and graphite, disappear in the spectrum from the oxygen-exposed sample. This may happen due to oxidation of carbon.

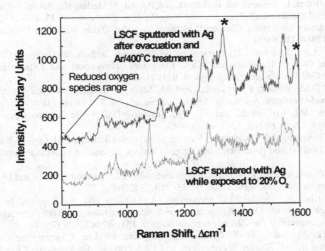

Figure 6. Raman spectra collected at room temperature from LSCF with sputtered Ag under controlled atmosphere conditions.

CONCLUSIONS

From the results of the experiments outlined above, these SERS methodologies have shown significant promise in characterizing of the surface structure and chemistry of SOFCs. In situ experiments with controlled atmospheres have been shown to yield pertinent information on key surface species for SOFC processes, such as the reduced oxygen that drives interfacial polarization. Future work in this field includes analyzing cathode samples with these SERS methodologies under various applied electrochemical conditions and environments closer to actual SOFC operating temperatures. The data collected in this investigation might be supplemented by simulation work, such as quantum chemical calculations, in order to complete an overall understanding of SOFC cathode surfaces.

ACKNOWLEDGMENT

This work was supported by DOE-NETL University Coal Program (Grant No. DE-FG26-06NT42735) and DOE Basic Energy Science (Grant No. DE-FG02-06ER15837).

REFERENCES

[1]J. Will, A. Mitterdorfer, C. Kleinlogel, D. Perednis, and L.J. Gauckler, Fabrication of Thin Electrolytes for Second-Generation Solid Oxide Fuel Cells, *Solid State Ionics*, **131**, 79-96 (2000).

[2]J.P.P. Huijsmans, F.P.F.v. Berkel, and G.M. Christie, Intermediate Temperature SOFC - a Promise of the 21st Century, *Journal of Power Sources*, **71**, 107-10 (1998).

[3]C. Hwang, C.-H. Tsai, C.-H. Lo, and C.-H. Sun, Plasma Sprayed Metal Supported YSZ/Ni-LSGM-LSFC ITSOFC with Nanostructured Anode, *Journal of Power Sources*, **180**, 132-42 (2008).

[4]J. Deseure, Y. Bultel, L. Dessemond, E. Siebert, and P. Ozil, Modelling the Porous Cathode of a Sofc: Oxygen Reduction Mechanism Effect, *Journal of Applied Electrochemistry*, **37**, 129-36 (2007).

[5]S.B. Adler, Factors Governing Oxygen Reduction in Solid Oxide Fuel Cell Cathodes, *Chemical Reviews*, **104**, 4791-843 (2004).

[6]C. Nicolella, A.P. Reverberl, P. Carpanese, M. Viviani, and A. Barbucci, Influence of the Temperature on Oxygen Reduction on SOFC Electrodes: Theoretical and Experimental Analysis, *Journal of Fuel Cell Science and Technology*, **5**, 011011-15 (2008).

[7]P.J. Hendra, I.D.M. Turner, E.J. Loader, and M. Stacey, The Laser Raman Spectra of Species Adsorbed on Oxide Surfaces, *Journal of Physical Chemistry*, **78**, 300-04 (1973).

[8]V.V. Pushkarev, V.I. Kovalchuk, and J.L. d'Itri, Probing Defect Sites on the CeO_2 Surface with Dioxygen, *J. Phys. Chem. B*, **108**, 5341-48 (2004).

[9]A. Zecchina, G. Spoto, and S. Coluccia, Surface Dioxygen Adducts on MgO-CoO Solid Solutions: Analogy with Cobalt-Based Homogeneous Oxygen Carriers, *Journal of Molecular Catalysis*, **14**, 351-55 (1982).

[10]D. Uy, A.E. O'Neill, and W.H. Weber, UV Raman Studies of Adsorbed Oxygen and NOx Species on Pt/Gamma-Alumina Catalysts, *Appl. Catal. B*, **35**, 219-25 (2002).

[11]T. Itoh, K. Abe, K. Dokko, M. Mohamedi, I. Uchida, and A. Kasuya, In Situ Raman Spectroelectrochemistry of Oxygen Species on Gold Electrodes in High Temperature Molten Carbonate Melts, *Journal of The Electrochemical Society*, **151**, A2042-A46 (2004).

[12]Y.M. Choi, H. Abernathy, H.-T. Chen, M.C. Lin, and M. Liu, Characterization of O_2-CeO_2 Interactions Using in Situ Raman Spectroscopy and First-Principle Calculations, *ChemPhysChem*, **7**, 1957-63 (2006).

[13]M.G. Albrecht and J.A. Creighton, Anomalously Intense Raman Spectra of Pyridine at a Silver Electrode, *J. Am. Chem. Soc.*, **99**, 5215-17 (1977).

[14]K. Kneipp, H. Kneipp, I. Itzkan, R.R. Dasari, and M.S. Feld, Ultrasensitive Chemical Analysis by Raman Spectroscopy, *Chemical Reviews*, **99**, 2957-75 (1999).
[15]Y. Liu, S. Zha, and M. Liu, Novel Nanostructured Electrodes for Solid Oxide Fuel Cells Fabricated by Combustion Chemical Vapor Deposition (CVD *Adv. Mater.*, **16**, 256-60 (2004).
[16]M. Pomfret, J.C. Owrutsky, and R.A. Walker, In Situ Studies of Fuel Oxidation in Solid Oxide Fuel Cells, *Analytical Chemistry*, **79**, 2367-72 (2007).

CHARACTERIZATION OF AN ANODE-SUPPORTED PLANAR SOLID OXIDE FUEL CELL WITH A POROSITY CONCENTRATION GRADIENT

Chung Min An[1], Jung-Hoon Song[2], Inyoung Kang[1], Nigel Sammes[1]
[1]Department of Metallurgical and Materials Engineering, Colorado School of Mines, Golden, Colorado, USA; [2]RIST, Pohang, Republic of Korea

ABSTRACT

A standard Ni/YSZ cermet tape-cast anode, an 8mol%YSZ electrolyte, and lanthanum strontium manganite (LSM) cathode were used for the fabrication process of a solid oxide fuel cell (SOFC) unit. An anode-supported electrolyte was prepared using a tape casting technique followed by hot pressing lamination and a single step co-firing process, allowing the creation of a thin layer of dense electrolyte. Scanning electron microscopy (SEM) revealed a crack-free and dense structure of the electrolyte in the unit cell. The unit cell exhibited good performance and demonstrated that the concentration distribution of porosity in the anode increases the power in the unit cell. Active layers, with different rates of pore former, were added in the anode slurry, and the different tapes (with different pore-formers) were then successfully laminated together.

INTRODUCTION

Fuel cells have been considered for many years as a possible solution for energy requirements with lower pollution. Many types of fuel cells have been investigated, including proton exchange membrane fuel cells (PEMFC), molten carbon fuel cells (MCFC), alkaline (AFC), and solid oxide fuel cells (SOFC)[1]. A planar-SOFC, which includes active anode layers with different porosities, is the focus of this paper. The SOFC is composed of an anode (Ni/8mol% Y_2O_3-stabilized ZrO_2) (Ni/YSZ), electrolyte (8mol% Y_2O_3-stabilized ZrO_2) (YSZ) and cathode ($La_{0.8}Sr_{0.2}MnO_3$) (LSM)[2, 3]. Due to the sintering temperature of the SOFC being over 1000°C, many physical and chemical problems are prevalent when fabricating them[4]. The focus of research in SOFC's is, thus, forced to the intermediate operating temperature below 800 °C. There are a number of ways to achieve this, including using a thin electrolyte layer (below 20μm); changing the electrolyte material in order to get a good ionic conductivity in the intermediate temperature range; finally, active layers can be added on the anode site to increase the triple phase boundary (TPB), and thus potential reactivity.

Methods for fabricating the anode-supported SOFC include electrochemical vapor deposition (EVD), chemical vapor deposition (CVD), screen printing, and tape casting[5, 6]. Tape casting is one technique that can be used for mass production of the SOFC, for instance, continuous fabrication processing, easily making a porosity gradient in the anode, and so on[7].

This research focuses on the effect of the porosity gradient. The process of fabricating green ceramics for the anode, electrolyte and active layer in the anode is via tape casting, while screen-printing is used for the cathode coating. Due to TPB being dependent on porosity, the performance is expected to be related to the porosity gradient, because the porosity concentration is dependent on the pore former concentration in the slurry.

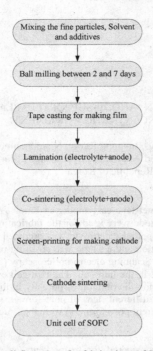

Figure 1. Overall flow chart for fabrications of SOFC unit cell

EXPERIMENT

In this study, 8mol% YSZ (TZ-8Y, Tosoh Co., Japan; FYT 13-010H, Unitech Ceramics, UK), NiO (Ferro), carbon black (Ravern 430, Columbian Chemicals, USA), modifier (M1201, Ferro Co., USA), binder (B74001, Ferro Co., USA) and solvent (ethanol and toluene) were mixed together to make the tape-cast slurry for the anode[3]. For fabricating the green film of the electrolyte using the tape casting technique, the 8mol% YSZ (TZ-8Y, Tosoh Co., Japan), the modifier (M1201, Ferro Co., USA), the binder (B74001, Ferro Co., USA) and the solvent (ethanol and toluene) were used. LSM (Fuelcellmaterials, LSM 20) was used for the cathode and, as explained below, was screen printed onto the electrolyte layer. The processing steps for making the unit cell are shown in the flow chart, Figure 1.

Figure 2 represents a pictorial view on how to produce the porosity gradient in the anode. The porosity is realized by the concentration of carbon black in any particular anode layer; four kinds of active layers are made, the concentration of carbon black in each layer being 0%, 0.94%, 1.86% and 2.76%. Each active layer thickness was between 10μm and 20μm after co-sintering.

To make the sample SOFC cells, which included the active layers, every ingredient was mixed using a ball milling process for two days. The thin films were fabricated using tape casting (HANSUNG SYSTEMS, STC-28A). Highly porous layers in the anode, in which the green ceramic thickness was 200μm before drying the tapes, were prepared. The active layer was prepared with a thickness of 90μm after drying the green ceramic film. After preparing the films, the green ceramic underwent a lamination process, depending on the type of SOFC system required (Fig. 2). The operating temperature of the lamination step was set from 80 to 100 °C while the pressure was gradually raised to 4000 psi. After forming the button cells by the lamination step, thus producing an anode/electrolyte couple, they were co-sintered at 1350 °C for 3 hrs.

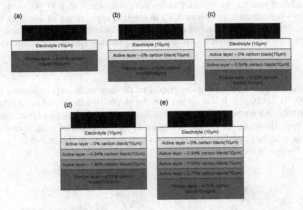

Figure 2. Schematics of unit cells

Figure 3. Unit cells (a) after reduction process (b) before coating cathode

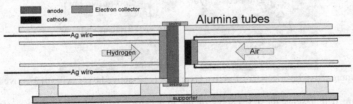

Figure 4. Test cell for the unit cells

The cathode was fabricated by screen printing with a paste of a commercial LSM product. After coating the cathode on the sintered anode/electrolyte couple, the cells were re-sintered at 1150 °C for 3 hrs. Finally, unit cells were fabricated with a diameter of the anode being 3cm, and the cathode being 1.5cm, as shown in Figure 3.

The test cell to measure the electrical performance is shown by figure 4. Current-voltage characteristics were measured using an electric load (Chroma System Solutions Inc. DC electronic load) at 800 °C, in 3% humidified hydrogen (50cc/min) for the anode, and in dry air (100cc/min) for the cathode. Impedance spectra of the unit cells were measured at 800 °C using AC impedance spectroscope (Gamry Instruments, reference 600) in the frequency range of 100KHz to 0.02Hz with a signal amplitude of 5mV under open circuit conditions. The microstructure of the unit cells, formed on the PET film by the tape casting process, and the sintered anode-supported electrolyte, were identified using SEM.

RESULTS AND DISCUSSION

Figure 5 shows cross-sectional micrographs of the unit cells after reduction in the hydrogen-environment at 800 °C for 10 hrs. It was found that the electrolyte layer had no cracks, and the anode layer had high porosity. As shown in Figure 2, the thickness of each active layer was expected to be approximately 10 μm. However, as shown in Figure 5, the thickness of each active layer in the unit cells, after sintering, was increased due to the increasing porosity in the active layer, as a function of the increased pore former in the slurry.

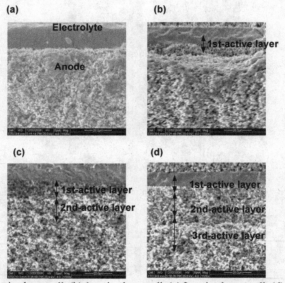

Figure 5. (a) non-active layer cell, (b) 1-active layer cell, (c) 2-active layers cell, (d) 3-active layers cell

Figure 6 represents the I-V curves of the unit cells. The electrochemical tests for the I-V curves and the impedance of the unit cells were measured at 800 °C. The OCV's of the cells were approximately 0.97 ~ 1.05V. When the number of active layers in the unit cells was increased, the IR drop across the cell, on drawing current, reduced. However, in the case of the 4-active layer cell, the IR drop was rather higher than those of the 1-, 2- and 3-active layer cells, this could be postulated as being due to too high an increase in the thickness of the anode and a potential loss in anode performance. From the I-V curves in Figure 6, the power output of the cells is 0.076 (for non-active layer), 0.086 (for 1-active layer), 0.097 (for 2-active layers), 0.101 (for 3-active layers), and 0.081 (for 4-active layers) W/cm^2. The 3-active layer cell was found to have the highest power output at 0.101 W/cm^2. However, the performance of the 4-active layers cell had decreased, which is most likely due to a larger porosity gradient[8]. Although the power output is not as high as other work, and other systems, the potential for using this technique is high. It is apparent that a porosity gradient can be fabricated across an anode by tape casting different porous anode layers, and then laminating them together, with favorable results. The process does, however, need to be optimized.

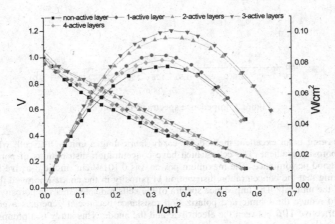

Figure 6. I-V and characterization of the unit cells

To examine the Ohmic resistance of the electrolyte and the polarization resistance of the electrode in more detail, impedance spectra were conducted under the same conditions as that for the electrochemical testing. Figure 7 shows the results of the test, in which Ohmic resistances were determined to be 0.88 and 0.84, 0.75, 0.73, 0.86Ωcm^2, and the polarization resistances were 0.96 and 0.91, 0.78, 0.78, 1.0 Ωcm^2, for the non, 1-, 2-, 3-, and 4- active layer-systems, respectively. As a result of the Ohmic and polarization losses, resistances of the 3-active layer cell is the lowest. Because of the porosity gradient, the TPB site would be increased and hydrogen could easily go to the TPB site through the porosity gradient. However, if the porosity gradient is extended, as shown in the 4-active layer cell, hydrogen gas diffusion to the TPB site would be more difficult; however the actual mechanism is still under scrutiny. Although the results of the unit cell show lower performance than is reported in the literature for similar systems, due to fabrication issue that need to be optimized, it can still be concluded that the introduction of a porosity gradient could increase its performance over other possible scenarios.

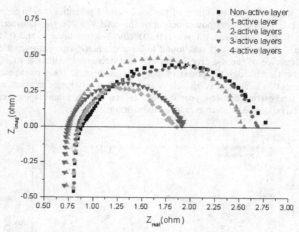

Figure 7. Impedance spectra of the unit cells

CONCLUSIONS

Tape casting is an excellent method for easily fabricating a unit SOFC cell, which has an anode with a porosity gradient. The cells which have concentration distributions of porosity in the anode showed good performance. The maximum power was $0.101 W/cm^2$ in the 3-active layer cell at 800 °C, indicating that the concentration distribution of porosity in the anode improved the electrical performance in the unit cell. The impedance spectra of the cells indicated that the porosity gradient in the anode could reduce the Ohmic and polarization resistance because it generates a good contact geometry and effective TPB between the electrolyte and the anode. This study is a promising one, in that the performance increased by having a concentration distribution of porosity in the anode; further work is underway to try and better understand these results from a kinetics point of view.

REFERENCES

[1]S.P.S. Badwal, K. Foger, Solid Oxide Electrolyte Fuel Cell Review, Ceramics International 22, 257-265 (1996)
[2]H. Moon, S. D. Kim, S. H. Hyun, H. S. Kim, Development of IT-SOFC unit cells with anode-supported thin electrolytes via tape casting and co-firing, International journal of hydrogen energy 33, 1758-1768 (2008)
[3]J. Song, S. Park, J. Lee, H. Kim, Fabrication characteristics of an anode-supported thin-film electrolyte fabricated by the tape casting method for IT-SOFC, Journal of materials processing technology 198, 414-418 (2008)
[4]S. D. Kim, S. H. Hyun, J. Moon, J. Kim, R. H. Song, Fabrication and characterization of anode-supported electrolyte thin films for intermediate temperature solid oxide fuel cells, Journal of Power Sources 139, 67-72 (2005)
[5]M. Inaba, A. Mineshige, T. Maeda, S. Nakanishi, T. Iomi, T. Takahashi, A. Tasaka, K. Kikuchi, Z. Ogumi, Growth rate of yttria-stabilized zirconia thin films formed by electrochemical vapour-deposition using NiO as an oxygen source, Solid State Ionics, 104, 303-310 (1997)
[6]X. Ge, X. Huang, Y. Zhang, Z. Lu, J. Xu, K. Chen, D. Dong, Z. Liu, J. Miao, W. Su, Screen-printed thin YSZ films used as electrolytes for solid oxide fuel cells, Journal of Power Sources, 159, 1048-

1050 (2006)

[7]D. Simwonis, H. Thulen, F.J. Dias, A. Naoumidis, D. Stover, Properties of Ni/YSZ porous cermets for SOFC anode substrates prepared by tape casting and coat-mix process, 'Journal of Materials Processing Technology, **92-93**, 107-111 (1999)

[8]K. Chen, X. chen, Z. Lu, N. Ai, X. Huang, W. Su, Performance evolution of NiO/yttria-stabilized zirconia anodes fabricated at different compaction pressures, Electrochimica Acta **54**, 1355-1361, 2009

IMPACT OF PROTECTIVE AND CONTACTING LAYERS ON THE LONG-TERM SOFC OPERATION

Mihails Kusnezoff[1], Stefan Megel[1], Viktar Sauchuk[1], Egle Girdauskaite[1], Wieland Beckert[1], Andreas Reinert[2]

[1] Fraunhofer Institute for Ceramic Technologies and Systems
Winterbergstr. 28, 01277 Dresden, Germany
[2] Staxera GmbH
Gasanstaltstr. 2, 01237 Dresden, Germany

ABSTRACT

Different combinations of protective and contacting layers are investigated to decrease a long-term degradation and high temperature oxidation of Fe-Cr alloys/steels. The protective coatings influence both factors which cause the degradation of the stack: growth of the resistance of the oxide scale and chromium release from interconnect. The porous contacting layer helps to decrease the contact resistance between the interconnect and the cathode and is applied as a rule on top of the protective layer or on the cathode surface. The oxidation of interconnect as well as interaction between interconnect and porous ceramic layers based on perovskite (LSMC) and spinel (MCF, CNM) materials is investigated. It was found that even porous coating applied to the interconnect generally reduces oxide scale growth and chromium release. On basis of developed method for measurement of oxide scale resistance the ASR values of oxide scale formed between ferritic alloy and ceramics as well as its increase with annealing time at 850° is investigated. It is found that the coating with spinel results in lower ASR values and degradation rates.

INTRODUCTION

For the planar design of the solid oxide fuel cell several different ceramic materials[1] and metals[2] have been evaluated as interconnects over the past ten years, however a satisfactory low cost solution has not been found.

An interconnect material should fulfill following criteria to be suitable for application in planar SOFC:

- No open porosity. The material must be impervious to gases to avoid mixing of air and fuel inside the stack.
- High electronic conductivity (>50 S/cm) to provide low losses of the current transport in the stack.
- Phase and long term stability in both oxidizing and reducing environments.
- Negligible ionic conductivity at the operation temperature (700-950°C) to avoid internal short circuiting between the air side and fuel side.
- Low thermal expansion mismatch with other components. The thermal expansion coefficient of the most other fuel components[3] varies between 10 and 13×10^{-6} K^{-1}.
- High thermal conductivity (=20 $Wm^{-1}K^{-1}$) for an uniform heat distribution in the stack.
- Reasonable strength because the interconnect material in planar stacks provides structural support.

Common metals and alloys have thermal expansion coefficients much higher than that of other SOFC components. They have high creep rates and corrode rapidly at the operation temperature of the common electrolyte supported cell (850-950°C). The first oxide dispersion strengthened alloy $Cr5Fe1Y_2O_3$ suitable for long term operation at temperatures up to 950°C were developed jointly by Plansee and Siemens[1]. The major advantages of this alloy are the high tensile strength, high thermal conductivity, good creep resistance and relatively low thermal expansion ($\alpha_{RT..1000°C}$=11.4 ppm/K) which matches with 8 mol% Y_2O_3-ZrO_2 ($\alpha_{RT..1000°C}$=10.8 ppm/K) in the whole temperature range. It has been shown by several authors[2,3] that the use of uncoated $Cr5Fe1Y_2O_3$ alloy cause the degradation of the cathode performance due to CrO_3 release and $(Cr,Mn)_3O_4$-spinel formation at the three phase boundaries in the $(La,Sr)MnO_3$-based cathode especially at high current densities[2]. To prevent the chromium release from the metallic interconnect protective layers based on $La_{1-x}Sr_xCrO_3$, $La_{1-x}Sr_xMnO_3$[4] and $Y_{1-x}Ca_xMnO_3$[5] were developed and used in combination with perovskite contact layers. Nevertheless the poor mechanical machinability of the $Cr5Fe1Y_2O_3$ and additional costs for protective coatings prevented their broad market penetration and only with introduction of new net-shape technology for manufacturing of interconnects the CFY material is on the way to be used in long-life SOFC stacks.

The novel concepts for the cell design using thin electrolyte layer on the anode or metallic support allowed lower SOFC operation temperatures (700-800°C) and the possibility to use ferritic steels as interconnect material.

The first commercial steels investigated and used for the stack manufacture were 1.4742[6] and the alloy 446 as well as SUS 430[7] and ZMG232[8]. The novel materials such as HZM-steels from FZJ[9] and Fe-26Cr-Ti-Y$_2$O$_3$ [10] have been intensively developed and leaded to commercial products such as Crofer22 and IT02-alloy respectively. Concurrently the commercial steels such as ZMG232 will be consequently modified [11].

These new steels are developed to fulfill two additional criteria at operating conditions:
– conductive and thin oxide layers with low growth kinetics
– low chromium release rates.

The most critical issue was the oxide scale formation in air due to a more rapid oxide growth and the principal difficulties to realize reliable electrical contact between cathode and metallic interconnector.

The degradation of the cathode performance as a result of its poisoning with chromium released from metallic interconnect is one of the major degradation mechanisms affecting the long-time stability of the SOFC stack. A chromium retention layer can be deposited onto interconnect to prevent a rapid deterioration of the cell parameters. The common strategy for the development of protection layers is the realization of the gas-tight films having a material with coefficient of thermal expansion (TEC) matched to the interconnect and a good electrical conductivity.

Different coating systems were tested in order to ascertain the most suitable SOFC interconnect protection material. Many approaches were based on the use of the layers of both single perovskite compounds (La,Sr)BO$_3$ (B=Mn, Co, Cr) and their combinations[12-17]. It was more recently established that the spinel compounds can also effectively reduce the Cr vaporization[18-20]. The (Mn,Cr)$_3$O$_4$ spinel phase is thermally formed in-situ in most interconnect alloys (Crofer22APU, ZMG232) via the formation of the outer spinel layer during the surface oxidation of the material. The in-situ formed double-layer oxide structure consisting of upper spinel layer und chromia under layer can only lower the Cr release by factor two compared to uncoated release rate. However, it is not good enough to ensure a long life time of the fuel cell since there is the chromium diffusion through the chromium-containing spinel scale. An application of the additional spinel containing protection layer is a more reliable method to prevent the chromium poisoning of the SOFC cathode. The spinel phases like Co$_x$Cu$_y$Mn$_{3-x-y}$O$_4$, Ni$_x$Cu$_y$Mn$_{3-x-y}$O$_4$, Co$_x$Fe$_y$Mn$_{3-x-y}$O$_4$, with 0≤x≤2, 0≤y≤1, and (x+y)<3 are particularly suitable as interconnect protection materials[21]. It was found that (Mn,Co)$_3$O$_4$ spinels have a good high temperature electrical conductivity (up to ~60 S/cm at 800°C)[22]. Together with the good thermal expansion match between (Mn,Co)$_3$O$_4$ and the metal substrate (thermal expansion coefficient (TEC): 13.4×10^{-6}K^{-1} for Mn$_{1.5}$Co$_{1.5}$O$_4$ vs. 12.4×10^{-6}K^{-1} for Crofer22APU[23]) these spinels are promising coating materials to improve the surface stability of ferritic stainless steel interconnects, minimize contact resistance and seal of chromium in the metal substrate.

In contrary to described effects the growth of oxide scale resistance due to interaction between air,alloy material and ceramic coating is not systematically investigated due to difficulties in reliable measurement of oxide scale resistance. The present work shows the impact of protective/contacting layers on the degradation rate of ASR due to oxide scale growth.

METHODS & EXPERIMENTALS

Two ferritic alloys Crofer22 and ITM (see Table I) are selected to demonstrate the impact of the oxide scale growth on the degradation rates of SOFC. The ceramic coatings (La,Sr)(Mn,Co)O$_3$ (LSMC), MnCo$_{1.9}$Fe$_{0.1}$O$_4$ (MCF) and Cu$_x$Ni$_{1-x}$Mn$_2$O$_4$ (CNM) for protecting of metallic component from chromium release and oxidation in air are investigated.

The composition and microstructure of the oxide layer as well as internal oxide precipitates in the metal matrix and the microstructure changes of the metal matrix were observed using field emission scanning electron microscope (Gemini 982, Fa. Leo) on the samples annealed for 800h and 2600h. Chemical composition of the samples was determined by an energy dispersive X-ray (EDX) analyzer and by

Table I   Chemical composition of selected ferritic alloys[24]

	Trademark of	Fe	Cr	Mn	Al	Si	Others
CroFer22APU	Thyssen-Krupp	bal.	22,0	0,42	0,12	0,11	La=0,08 Ti=0,08
ITM	Plansee	bal.	26,0	pres.	<0,03	<0,03	$(Mo)_x$, $(Ti)_y$, $(Mn)_z$, $(Y)_{xy}$

electron probe microanalysis using both point microanalysis and X-ray maps of the element distribution. The phases of the oxide scales were analyzed by reflection X-ray diffraction (XRD) using CuK_α radiation (XRD7, Fa. Seifert).

All samples were annealed at 850°C in air. The weight of all samples was measured using scales (Fa. Sartorius) before and after the annealing. The weight rise normed to the sample surface was determined to provide the merit for the oxide growth.

The four point electrical resistance of the oxide scales was studied at 850°C as a function of the annealing time. The samples with special geometry (Fig. 1) were used for measurement of electrical resistance of oxide scale. The three ribs made of ceramic material were dispensed on the surface of the metal sheet, slightly dried and contacted with similar sheet. The specific area resistance between alloy and contact rib is calculated using 3D simulation. To calculate this resistance following data were estimated in separate experiments and used in the simulation for further calculations: resistivity of rib material and of alloy, contact area and cross section area for current transport[29].

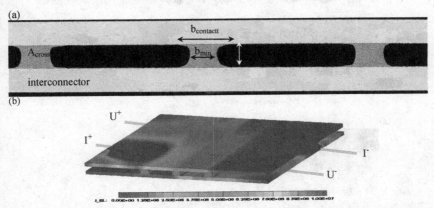

(a)

(b)

Fig. 1   a) Two metallic sheets contacted by ribs of ceramic material; b) current distribution in the sample calculated using 3D simulation (average current density 100 mA/cm²)

RESULTS & DISCUSSION
a. Oxide scale growth

Morphological observations and results of XRD analysis reveal that oxide scales formed on Crofer22 and IT02 alloys after thermal treatment for 800h consist of Cr_2O_3 with $MnCr_2O_4$-spinel and of Cr_2O_3 in the outer part of the scale respectively (Fig. 2). These oxides are electrically conductive and chemical compatible to the contact layers made from obviously used perovskites (especially manganates and chromites). After annealing of Mn-containing alloy at 850°C in air (see Fig. 2(a)) a double layer oxide scale is formed. At the outer oxide scale a $(Cr,Mn)_3O_4$-spinel (identified as $CrMn_{1.5}O_4$ phase from XRD peaks) on top of inner Cr_2O_3 layer is grown.

The formation of $(Cr,Mn)_3O_4$ layer on the surface of Cr_2O_3 is typical for Mn containing ferritic alloys due to the high diffusion coefficients of Mn in Cr_2O_3. Additional ZrO_2 and TiO_2 are present at the interface area of Crofer22APU. Pores (10-15 µm) decorated with SiO_2 and Al_2O_3 are main constituents of the sub- surface region and influence the adhesion of the oxide scale. The metallic matrix is affected by micro-pores and precipitations along the grain boundaries. The grains in the annealed bulk material are coarsened in comparison to the initial state because of recrystallization. The recrystallization at the sub-surface region is inhibited by the precipitions at the grain boundaries.

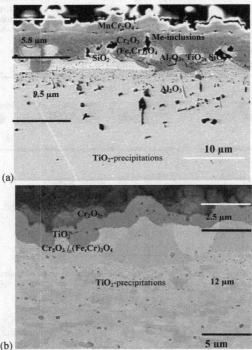

Fig. 2   Oxide scale after annealing for 800h at 850°C for Crofer22APU (a) and ITM (b).

ITM has the thinnest oxide scale and corroded sub-surface area. The oxide scale consists of Cr_2O_3. The pores have been found not only at the oxide scale / metal interface but also in the Cr_2O_3 layer of the oxide scale. In the sub-surface region the precipitations of TiO_2 are found.
Existence of precipitations in sub-surface area in both materials clearly shows that the oxide scale is permeable for oxygen and possibly has voids and cracks.

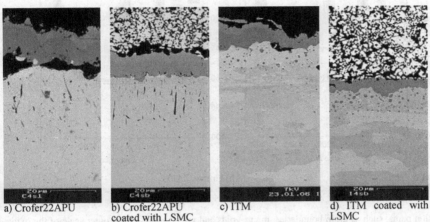

a) Crofer22APU   b) Crofer22APU   c) ITM   d) ITM coated with
              coated with LSMC   LSMC

Fig. 3   Oxide scale growth after annealing in air at 850°C for 6400h

Generally every coating of alloy surface significantly reduces the oxidation rate of interconnect material (see Fig. 3). Fig. 4 shows the oxide scale thickness as a function of annealing time. Based on obtained data the prediction for breakaway oxidation of uncoated and coated interconnects with the thickness of 0.5 mm is made. Because the oxide scale at the air side is thicker than at the fuel side, an oxide scale growth from both surfaces of the interconnect in air would be the worst case.
The weight gain is caused not only by the growth of the outer oxide scale but also through the formation of precipitates in the metallic matrix. For both materials the thickness of the oxide layer and corroded sub-layer increases with the annealing time. The weight gain after annealing of the investigated steels at 850°C is measured (Fig. 4). The weight change for ITM samples up to 400h was too small and could not be properly estimated by the used method.
If the oxide scale growth proceeds by diffusion of metal and/or oxygen ions through the oxide lattice, growth rates will obey to a parabolic time dependence in coincidence with the classical Wagner theory. The weight gain for Crofer22 does not obey to the parabolic law because the assumptions made in Wagner theory to describe the scale growth are not fulfilled. Oxide scale growth proceeds via rapid

Table II Estimated lifetime for different alloys with initial thickness of 0.5mm

Alloy	Approximation for oxide scale growth in air (uncoated)	Lifetime (uncoated) [hours]	Approximation for oxide scale growth in air (coated)	Lifetime (coated) [hours]
CroFer22APU	x = 0,005 t+32	13.600	x = 0,0033 t+35	19.700
ITM	x = 0,0034 t+19	25.900	x = 0,0024 t+9	37.900

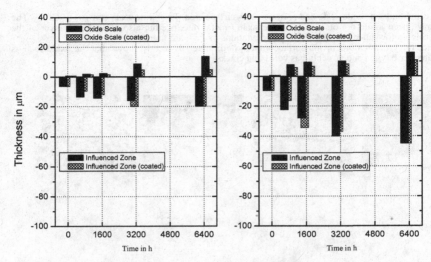

a) ITM   b) Crofer22APU
Fig. 4   Thickness of oxide scale and influenced (sub-surface) zone of alloys annealed at 850°C
          in air

diffusion paths, such as grain boundaries, micro-voids and micro-cracks in the outer scale of the oxide layer. Furthermore the formation of volatile oxides can affect chromia, alumina, silica and spinel growth at high temperatures as well as weight changes of the sample.

The analysis of the thickness of oxide scale is more reliable parameter for description of the oxidation behaviour. For lifetime calculations the thicknesses of outer and inner oxide scales were added and plotted versus time. Calculated lifetime for 0.5 mm thick interconnector assuming the allowable remaining uncorroded thickness of 0.3 mm (60% of specimen thickness) is shown in Table II.

The used linear approximations are not reflecting the model of Wagner but are a better fit to the measured oxide scales after anneling at 850°C for 1600, 3200 and 6400 hours.

The estimated life time depends on the initial thickness of the specimens. The critical thickness of non-oxidized material has to be found out experimentally in further investigations for better predictions of stack life time defined by brakeaway oxidation.

The coating with porous ceramic layers of LSMC, MCF and CNM does not change principal structure of the oxide scale. However the interaction and composition of the outer oxide scale is strongly affected by the ceramic material used (see Fig. 5).

The LSMC protective layer is well connected to the Crofer22 surface (Fig. 5a). Only small interdiffusion of Co and strong interdiffusion of Mn takes place resulting probably in thicker $(Cr,Mn)_3O_4$ spinel layer on top of the Cr_2O_3 in comparison to the uncoated substrate. However the overall thickness of oxide scale is less for samples with LSMC coating (see Fig. 3).

The cross-sectional image (Fig. 5b) of the samples after the oxidizing heat-treatment shows that the MCF protection layer is well bonded to the Crofer22APU substrate via a thin scale interlayer which was grown between the spinel protection layer and the metal substrate. The elemental analysis of the interlayer shows that it is mainly composed of oxidized Cr and Mn.

Fig. 5c shows the SEM images of the cross-section of the roll-coated CNM protection layer on the Crofer22APU substrate after exposure to air at 850°C. The protection layer is bonded to the substrate via the typical scale consisting of the double-layer oxide structure. However the outer layer of the oxide scale has now the composition of (Mn,Cr,Cu,Ni)-oxide. Such a structure of the outer oxide scale layer

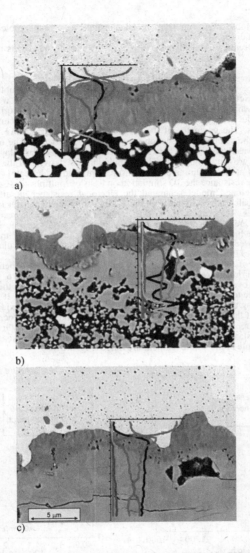

Fig. 5 Oxide scale after annealing for 1600h at 850°C for Crofer22APU with different coatings: (a) LSMC, (b) MCF and (c) CNM. Inserted EDX line-scan with O, black line; Cr, red line; Mn, blue line; Fe, olive line; Co, magenta line; Sr, purple line; La, cyan line; Ni, brown line; Cu, green line.

provides a more smooth transition from metal substrate to ceramic protection layer damping the mismatches between different layers. The layers are still porous after the heat-treatment. However the grainy structure inherent to non-sintered particles of the starting powder is not more observed in the layers.

b. Electrical resistance of oxide scale and impact on degradation

The resistance value measured by four point method for the sample in Fig. 1 includes the resistances of ceramic layer, oxide scale and metallic interconnector as well as is affected by the geometrical factors which varies from sample to sample during manufacturing. The temperature and time dependent conductivity of ceramic layers and of metallic interconnector have been estimated in separate experiments. The geometry of the ribs and the contact area have been estimated by post-mortem analysis of the sample cross-section using optical microscopy. The only unknown part of the total resistance left was the resistance between interconnector and ceramic layer which is equivalent to the resistance of the growing oxide scale. The simulation has shown that the current distribution over the sample especially at the beginning of the measurement (thin oxide scale) exists. To provide reliable values for oxide scale resistance the 3D simulation on basis of Multiphysics-Code ANSYS 5.6 has been performed. The developed model separated the electrical resistance in different constituents. The resistance of oxide scales can be found from the 3D model by the parameter inversion and comparison with measured values. The results of the estimation of the oxide scale resistance using the applied inversion algorithm are demonstrated on the example of LSMC ceramic layer. Three ribs of the manufactured sample have the average width of 1,488 mm, hight of 0,637 mm and length of 19 mm. The contact area between ceramic rib and the interconnector surface is calculated using average values for width and length of the ribs.

The total resistance of the sample is shown in the Fig. 6. It is time dependent and increases with annealing time. The impact of interconnector (specific resistance 0,116 mΩcm^2), ceramic layer (resistance decreases over the annealing time due to sintering of particles in the porous layer) and oxide scale resistance on the total resistance of the sample are shown in Fig. 6(a). It is clearly seen that at the beginning of annealing the ceramic ribs resistance is even higher than the oxide scale resistance (due to the small thickness of the oxide scale at the beginning of the annealing). Proceeding the time the oxide scale grows and the impact of the oxide scale resistance increases. After 700h the values of oxide scale and ceramic rib resistance are equal and after 1340 h the oxide scale resistance dominates the total resistance of the sample in the whole temperature range (Fig. 6(b)).

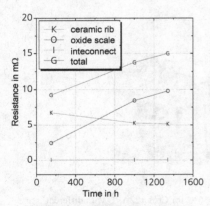 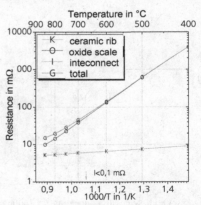

a) the constituents of total resistance as a function of time

b) the constituents of the total resistance after 1340 h / 850°C as function of temperature

Fig. 6   The constituents of the total resistance of sample from Fig. 1 obtained using 3D simulation.

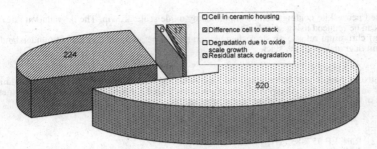

Fig. 7   Losses in the stack and degradation rates on example of two cell stack with LSMC ceramic ribs.

The estimation of oxide scale resistance for different samples with the same LSMC ceramic layer normalized to the contacting area gives ASR values in range of 1-1,3 mΩcm^2 and 4,2-5,0 mΩcm^2 after annealing for 150h and 1340h at 850°C respectively (only annealing time, the heating up and down time is not included). It means that the average of ASR increases from 150 to 1340h of about 2,8 mΩcm^2/1000h. This value can be good correlated to the degradation observed in the short stack experiments with LSMC contact ribs for 1000h of operation at 850°C.

Due to some distance between the contacting ribs, the roughness of cathode and ribs surface and the tolerances in the stack manufacturing the real contact area between interconnector and MEA varies between 30-60% of the total available electrode area depending on the ribs geometry. Overall ASR of repetitive unit in the stack was found to be 744 mΩcm^2. The overall increase of ASR during operation was about 23 mΩcm^2 (3,1% per 1000h). Fig. 7 shows the estimated distribution of losses in the stack as well as impact of the oxide scale growth (at least 6 mΩcm^2 can be attributed to this process taking into account real contact area in the stack) on the overall degradation. The oxide scale growth in this case causes at least 25% of overall degradation (0,75% per 1000h). .

The residual degradation (ca. 1,5-2,25% per 1000h) is attributed to other processes. Especially chromium release can be responsible for further degradation effects. However the chromium release rate up to now has not been quantitatively correlated to the degradation rates of the stacks. The application of dense ceramic layers (LSMC, MCF or CNM) would prevent the chromium release. Also application of porous coatings with getter phase for volatile chromium species (such as MnO$_x$, CoO$_x$ and others) can provide a good protection of cell against chromium poisoning.

CONCLUSIONS

The investigations of different ferritic interconnects show, that the life time of at least 6.000 hours at operating temperature of 850°C in air can be achieved also by using uncoated ferritic steel such as Crofer22APU. The superior ITM alloys can be further used for extended life applications such as stationary combined heat and power plants. The possible problem for all alloys is the spall off of the Cr$_2$O$_3$ oxide layer during operation or/and thermal cycling.

The porous perovskite coating reduces the growth of oxide scales and slows down significantly the breakaway oxidation. The chromia release is probably the main restriction why the dense coatings should be applied. It was found that the chemical interaction between pre-oxidized alloy and LSMC, MCF and CNM ceramic layers is minimal. The presence of Mn in the coating layer favour the formation of (Cr,Mn)$_3$O$_4$ spinel phase in the oxide scale.

The electrical resistance of the oxide scale between interconnector and ceramic coating can be estimated using special sample geometry and 3D simulation. The oxide scale dominates the resistance of contacting element after 1300h of operation at 850°C. The degradation rates of 0,75-1,5% per 1000h

for LSM based perovskite coatings can be explained by oxide scale growth. The degradation rates and ASR values can be reduced using spinel coatings.
The impact of chromium release on stack degradation should be analysed in details to understand the total long-term degradation of the stack.

ACKNOWLEDGEMENTS
The Saxony Ministry of Science and the Fine Arts (SMWK), Staxera GmbH and State Ministry for Economy and Technology (BMWi) are greatly acknowledged for funding the research activities on development of different ceramic coatings and testing of interconnect materials.

REFERENCES
[1] H. Greiner, E. Ralf, EP 0578855 B1, 1992.
[2] S.P.S.Badwal, R. Deller, K. Foger, Y Ramprakash, J.P Zhang, SSI 99 (1997) 297
[3] C. Günther, Ph.D. Thesis, University Erlangen-Nurnberg, Erlangen (1997)
[4] Y. Larring and T. Norby, J. of the Electrochem. Soc., 147 (9) (2000) 3251
[5] A. Plass, E. Batawi, W. Straub, K. Honegger and R. Diethelm in: A.J. McEvoy (Ed.), Proceedings of the 4th European SOFC Forum, Lucern/Switzerland, (2000) 889
[6] F. Tietz, D. Simwonis, P. Batflsky, U. Diekmann and D. Stöver in: K. Nisancioglu (ed.), 12th IEA Workshop "Materials and Mechanisms", Annex VII, Wadal / Norway (1999) 1[11] R.A. Brooker, S.C. Kohn, J.R. Holloway and P.F. McMillan, Structural controls on the solubility of CO_2 in silicate melts Part I: bulk solubility data, Chemical Geology, 2001, **174**, 225–239
[7] T. Brylewski, M. Nanko, T. Maruyama and K. Przybylski, SSI, 143 (2001) 131
[8] T. Uehara, T. Ohno, A. Toji in: J. Huijsmans (Ed.), Proceedings of the 5th European SOFC Forum, Lucern/Switzerland, (2002) 281
[9] K. Honegger, R. Diethelm and W. Glatz in: H. Yokokawa and S.C. Singhal (eds), SOFC VII, PV 2001-16, The Electrochem. Soc. Proceeding Series (2001) 803
[10] T. Uehara, A. Toji, K. Inoe, M. Yamaguchi, T. Ohno in: M. Dokya and S.C. Singhal (eds), SOFC VIII, PV 2003-07, The Electrochem. Soc. Proc. Series (2003) 914
[11] W.J. Quadakkers, H. Greiner, M. Hänsel, A. Pattanaik, A.S. Khanna, W. Malléner, Compatibility of perovskite contact layers between cathode and metallic interconnector plates. Solid State Ionics, 91 (1996), 55
[12] K. Fujita, K. Ogasawara, Y. Matsuzaki, T. Sakurai, Prevention of SOFC cathode degradation in contact with Cr-conataining alloy. J. Pow. Sourc., 131 (2004), 261.
[13] N. Orlovskaya, A. Coratolo, C. Johnson, R. Gemmer, Structural characterization of lanthanum chromite perovskite coating deposited by magnetron sputtering on an iron-based chromium-containing alloy as a promising interconnect material for SOFCs. J. Am. Cer. Soc., 87 (2004), 1981.
[14] Z. Lu, J. Zhu, Y. Pan, N. Wu, A. Ignatiev, Improved oxidation resistance of a nanocrystalline chromite coated derritic stainless steel. . J. Pow. Sourc., 178 (2008), 282.
[15] N. Pryds, B. Toftmann, J. Schou, P.V. Hendriksen, S. Linderoth, Electrical and structural properties of $La_{0.8}Sr_{0.2}Mn_{0.5}Co_{0.5}O_{3\pm\delta}$ films produced by pulsed laser deposition. Appl. Surf. Sci., 247 (2005), 466.
[16] M. Stanislowski, J. Frojtzheim, L. Niewolak, W.J. Quadakkers, K. Hilpert, T. Markus, L. Singheiser, Reduction of chromium vaporization from SOFC interconnectors by highly effective coatings. J. Pow. Sourc., 164 (2007), 578.
[17] Zh. Yang, J.S. Hardy, M.S. Walker, G. Xia, S.P. Simner, J.W. Stevenson, Structure and conductivity of thermally grown scales on ferritic Fe-Cr-Mn steel for SOFC interconnect applications. J. Electrochem. Soc., 151 (2004), A1825.
[18] S. Geng, J. Zhu, M.P. Brady, H.U. Anderson, X-D. Zhou, Zh. Yang, A low-Cr metallic interconnect for intermediate-temperature solid oxide fuel cells. J. Pow. Sourc., 172 (2007), 775.
[19] M. Bertoldi, T. Zandonella, D. Montinaro, V.M. Sglavo, A. Fossati, A. Lavacchi, C. Giolli, U. Bardi, Protective Coatings of Metallic Interconnects for IT-SOFC Application. J. Fuel Cell Sci. Tech., 5 (2008), 011001.
[20] M. Zahid, F. Tietz, F.J.P. Abellon, Verfahren zur Herstellung einer Chromverdampfungschicht für chromoxidbildende Metallsubstrate. Patent DE 10 2005 015 755 A1, 12.10.2006.
[21] Zh. Yang, G-G. Xia, X-H. Li, J.W. Stevenson, $(Mn,Co)_3O_4$ spinel coatings on ferritic stainless steels for SOFC interconnect applications. Int. J. Hydr. Energy, 32 (2006), 3648.

[22] Zh. Yang, G. Xia, S.P. Simner, J.W. Stewenson, Thermal Growth and Performance of Manganese Cobaltite Spinel Protection Layers on Ferritic Stainless Steel SOFC Interconnects. J. Electrochem. Soc., 152, 9 (2005), A1896.
[23] W. Glatz, G. Kunschert, M. Janousek, and A. Venskutonis, P/M processing and properties of high performance interconnect materials and components for SOFC applications. Proceedings of the International Symposium SOFC-IX, Quebec, Canada, S. C. Singhal and J.Mizusaki, Editors, Pennington, p. 1773 (2005)
[24] S. Megel, Ph.D. Thesis, Technical University Dresden (in preparation)

CURVATURE EVOLUTION AND CONTROL IN ANODE SUPPORTED SOLID OXIDE FUEL CELLS

Marco Cologna, Anna Rita Contino, Vincenzo M. Sglavo
DIMTI, Università degli Studi di Trento
Trento, Italy
INSTM, Trento Research Unit
Firenze, Italy

Stefano Modena, Sergio Ceschini, Massimo Bertoldi
SOFCpower S.r.l.
Mezzolombardo (TN), Italy

ABSTRACT

In the present work anode supported half-cells with yttria (8 mol%) stabilised zirconia (YSZ) electrolyte and YSZ/NiO anode were fabricated by tape casting and co-sintering. In situ high temperature observations were carried out by a specifically designed microscope system in order to determine the curvature evolution of the cells upon firing. Such analytical observations together with careful tailoring of the starting powder allowed to drastically diminish the overall sintering stresses mismatch and related deficiencies. The developed microstructures observed by scanning electron microscopy confirmed the good quality of the obtained cell and in particular of the thin electrolyte layer.

INTRODUCTION

Solid oxide fuel cells (SOFC) are devices constituted of different layers each of them possessing different physical, chemical and thermal properties, which are sintered or co-sintered to form a multilayered structure.[1-3] In general, co-sintering of ceramics generates stresses because of different sintering kinetics and thermo-elastic properties of the constituting layers. In particular, the sintering stresses mismatch is of the same order of magnitude of the free sintering stresses, which in turn can be of the order of few megapascals. Such stresses are scaling inversely with grain size; nevertheless, in order to keep the sintering temperature to an acceptable level and to develop a sufficient fine electrode microstructure, SOFCs are generally fabricated starting from submicron sized powder and in some case from nano-sized powders. The elevated stresses which are developed upon co-sintering of fine powders compacts can reduce the effect of the sintering pressure and hinder densification, and in the worst cases can induce crack nucleation and propagation.[4-11] In the simple case of a bi-layer, such as a half-cell consisting of a single anode layer and an electrolyte, a curvature is generally developed upon the sintering process due to mismatch in shrinkage rates. The normalized curvature rate (dk/dt) is proportional to the mismatch in shrinkage rates and can be expressed as:[12, 13]

$$\frac{dk}{dt} = \dot{k} = \frac{6(m+1)^2 mn}{m^4 n^2 + 2mn(2m^2 + 3m + 2) + 1} \Delta \dot{\varepsilon} \tag{1}$$

where $k = (t_1 + t_2)/r$ is the normalized curvature, t_1 and t_2 are the layers thickness, r the curvature radius, $m = t_1/t_2$ the layer thickness ratio, $n = \eta_1/\eta_2$ the viscosity ratio in case of beam geometry, η being the uniaxial viscosity. The maximum tensile stress is developed in the fastest shrinking layer and is given by:

$$\sigma^{max} = \frac{m^2 n(4m+3) + 1}{m^4 n^2 + 2mn(2m^2 + 3m + 2) + 1} \eta \Delta \dot{\varepsilon} \tag{2}$$

95

Various approaches have been used in order to solve the problem of the curvature developed upon sintering, *e.g.* firing under a dead-load in a one-stage process,[14-16] performing a second creep flattening process after sintering,[17] or using a loading plate sustained by ceramic arches which yield at high temperatures.[18] Nevertheless, even if on a macroscopic scale such methods reduce the problem of curvature, on the microscopic scale they do not address the problem of crack growth and delamination which can be caused by sintering rate mismatch. Moreover, they convey various disadvantages including contamination from the substrate and increased friction, which can lead to failure or increase in the sintering anisotropy.[19]

From Eqs (1) and (2) one can appreciate that the developed curvature and stresses are dependent upon geometrical parameters (thickness ratio) and material properties (viscosity ratio). However, both ratios are fixed in the cell design. It is therefore crucial to reduce the anode and electrolyte sintering rate mismatch in order to reduce the developed stresses. The aim of the present work is to study the effect of different YSZ and NiO powder size and pre-conditioning on the sintering behavior of the resulting green tape and, therefore, on the curvature developed upon co-sintering of solid oxide half cells, with the final goal of minimizing the sintering stresses mismatch in order to obtain defect free and flat cells, eventually without the need of an external load.

EXPERIMENTALS

Green half cells (anode and electrolyte) were prepared by a sequential water-based tape casting technology, which consists in casting a thin electrolyte and then the support anode on top of the dried electrolyte tape. The electrolytes were cast with a blade gap of 30 μm and a carrier speed of 1.8 m/min; the anodes were cast with a blade gap of 700 μm and a carrier speed of 0.5 m/min. Slurry compositions are reported in Table 1. Fine (d_{50} = 0.3 μm, TZ8-YS, Tosoh, Japan) and coarse (d_{50} = 0.7 μm, FYT13-002H, Unitec, UK) 8 mol% yttria stabilised zirconia powders were used for the electrolyte. A mixture of 58 wt% NiO (J. T. Baker, USA, d_{50} < 1 μm) and 42 wt% TZ8-YS was selected for the anode. NiO was used in the form "as received" or calcined for 10 h at 900°C, ground in a mortar and sieved through a 100 μm mesh. The specific surface area (SSA) for the as received and calcined NiO was determined by nitrogen adsorption (BET) method (ASAP 2010, Micromeritics, USA). The microstructure of the produced materials was analyzed by scanning electron microscopy (SEM, JSM 5500, JEOL, Japan).

In a previous work[20] it was seen that a slow heating rate reduces the curvature rate in anode supported half cell. Therefore, in the present work a slow heating ramp (1°C/min) from 1100°C to 1450°C was chosen. Samples were sintered in a tubular furnace (HTRH 100-300/18, GERO Hochtemperaturöfen GmbH, Germany) up to 1450°C for 4 h. The shrinkage of monolithic layers and the curvature of half-cell bars (around 10 mm x 2 mm) were determined with a CCD camera acquiring pictures through a series of optical filters. Images were analyzed with ImageJ 1.38J freeware software. In case of curvature measurements, the electrolyte was always the top layer, while the anode was in contact with the support. A positive curvature was arbitrarily chosen when the electrolyte was on the concave side of the cell.

An LSM/YSZ cathode was screen printed on selected samples and electrochemical tests were performed as described in detail elsewhere[21] after sintering.

RESULTS AND DISCUSSION

SEM micrographs of the selected powders are reported in Fig. 1. Some differences appear clear. As received NiO (Fig. 1a) is strongly agglomerated and appears much finer compared to calcined NiO (Fig. 1b). The measured specific surface area is 3.3 m^2/g for the first, 1.6 m^2/g for the latter; the thermal treatment at 900°C is therefore significantly decreasing the specific surface area. Coarse YSZ particles (Fig. 1c) possess an irregular shape, while fine YSZ particles (Fig. 1d) are more regular and round shaped, being produced by a chemical route.

The four analyzed tapes are: two electrolytes - (i) Ec (prepared from the coarse YSZ powders), (ii) Ef (prepared from the fine YSZ powders) -, and two anodes - (iii) Ac (composed of fine zirconia and calcined NiO) and (iv) Af (composed of fine zirconia and "as received" NiO) -. True free strain, $\varepsilon = \ln (l(t)/l_0)$ and strain rate, $d\varepsilon/dt$, $l(t)$ being the instantaneous length as a function of time (t) and l_0 the initial length, are reported in Figs 2 and 3, respectively. In order to make comparison among different tapes easier, some reference data are reported in Table 2. As expected from the powder characteristics, electrolyte Ef is sintering at lower temperatures and with higher sintering rates compared to Ec. Comparing Ac and Af anodes, the first possesses a lower sintering rate. The coarsening treatment on NiO powders has therefore the effect of significantly diminishing the sintering rate. Comparing anodes and electrolytes sintering rates, the most critical combination with the highest mismatch in the sintering rate is Ac with Ef (0.64×10^{-3} min^{-1} at 1380°C, where Ec shows a linear shrinkage of 14% and Af of 7%, only); conversely, the best combination is Ac with Ec, the sintering rate mismatch being always lower than 0.1×10^{-3} min^{-1} over all the temperature range.

The developed normalized curvature $k = t/r$ (t being the bi-layer thickness and r the radius of curvature) and the cells curvature rate are reported in Figs 4 and 5, respectively. Both samples in which electrolyte Ef is used (Ac/Ef and Af/Ef) are bending towards the electrolyte side (positive curvature rate) in the first stages of sintering, as the electrolyte shrinks faster than the anode. The normalized curvature is reaching a maximum value (0.095 for Ac/Ef and 0.063 for Af/Ef) after which the curvature rate is inverting its sign. Sample Af/Ef is recovering an almost flat configuration at the end of the sintering process, while Ac/Ef remains strongly curved. The temperatures of curvature rate inversion, which corresponds to a point of zero mismatch stress, relate with a fairly good approximation to the point where the anode free sintering rate equals the electrolyte free sintering rate, both being in the range 1410°C - 1430°C.

On the other hand, sample Af/Ec is bending towards the anode side (negative curvature rate) in the firsts stages of sintering. The normalized curvature remains always negative and is maximal at the end of the sintering (-0.024). Only sample Ac/Ec possesses always a very low curvature rate and remains fairly flat over all the temperature range, being noticeably flat at the end of the sintering.

A positive curvature rate is the most critical condition for anode supported cell sintering process, since the thin electrolyte is subjected to tensile stresses, which retard densification, especially in the firsts stage of sintering, where ceramics possess still a limited strength and stresses may induce crack nucleation and propagation. A negative curvature rate is less critical, the electrolyte being under compression. As a consequence, the most critical sample is Ac/Ef, showing the maximum normalized curvature rate of 0.6×10^{-3}, followed by Af/Ec (dk/dt$_{max}$ = 0.5×10^{-3}). Af/Ec and Ac/Ec possess always a negligible positive curvature rate, one order of magnitude less than the maximum showed by Ac/Ef. Only Ac/Ec has always a very low positive or negative curvature rate, approximating very well the ideal conditions of a completely stress free co-sintering. The curvature trend of the half-cells is in very good agreement with the monolithic layers free sintering data: the cell built with the anode and electrolyte possessing the highest mismatch sintering rate (Ac and Ef) is the one showing the highest curvature rate, while the sample realized with Ac and Ec (mismatch sintering rate always lower than 0.1×10^{-3} min^{-1}) is almost flat over all the temperature range. The coarsening treatment on the NiO powders, by reducing the anode sintering rate, has a detrimental effect on the half-cell built using the fine electrolyte, while it is highly beneficial when the coarse electrolyte is used in the half-cell production, resulting in a virtually unconstrained sintering.

The cross section of as sintered Ac/Ec and Af/Ec samples is reported in Fig. 6. The electrolyte is retaining limited porosity, which is however not interconnected. No signs of delamination are observed. Anode Ac has clearly higher porosity compared to Af. The electrolyte surface of selected samples is shown in Fig. 7. The electrolyte of Af/Ec and Ac/Ec (Fig. 7a) is well densified and no cracks are visible at a careful examination. Flaws of tenths of micrometers in length and a few micrometers in width are clearly present on the electrolyte of Af/Ef (Fig. 7b). Crack of few hundreds of micrometers in length and tenths of micrometers in width are present on Ac/Ef (Fig. 7c and 7d) and are even visible at naked eye.

It is interesting to note that Ac/Ef, which possesses the highest mismatch in free strain rate between anode and electrolyte (Fig. 4), is developing the highest mismatch sintering stresses, which is leading to the development of the highest curvature rate (Fig. 5), crack extension and crack density (Fig. 7c and 7d). The only samples which are not showing any sintering defects are those showing negative or very limited curvature rate in the initial stages of sintering (anode sintering faster than electrolyte, like in Af/Ec, or anode sintering with the same rate as the electrolyte, like in Ac/Ec). It is worth nothing that for sample Ac/Ec, the anode and electrolyte sintering rate is so well matched, that the resulting half-cell is almost flat during the whole sintering cycle (Fig. 4), and perfectly flat at the end of the process. The very low curvature rate, which is observed in the intermediate stages of sintering (Fig. 5), is a clear indication of the very low stresses mismatch as generated upon sintering. This is a double advantage: first, a flaw free electrolyte and a perfect adhesion between anode and electrolyte can be obtained, as demonstrated by SEM micrographs; second, it allows the direct production of flat SOFCs without any creep flattening step, eliminating all the inherent disadvantages of such process.

By observing Fig. 7 it is clear that Af/Ef and Ac/Ef are not suitable for fuel cells production, since the cracks in the electrolyte would lead to fuel and oxygen mixing, while Af/Ac shows a pronounced curvature for being used without an additional flattening step; for these reasons, an electrochemical test was performed on Ac/Ec only. The data are plotted in Fig. 8. The power densities at 0.7 V at 750°C, 800°C and 850°C are 0.40, 0.49 and 0.62 Wcm^{-2}, respectively. Such performances, although lower than usually measured with samples under the same condition, have to be interpreted by noting that the anode microstructure was not optimized for electrochemical performance maximization. A thin denser anode functional layer (e. g. with the same composition of Af) may be easily inserted between the anode and the electrolyte in order to enhance the triple phase boundaries and the electrochemical performances, without varying the presented procedure and analysis.

CONCLUSIONS

Anode supported half-cells were produced by water-base sequential casting technology. The green tapes microstructure was varied and shrinkage and curvature analysis were performed in order to minimize the differences in the shrinkage kinetics between anode and electrolyte. An almost perfect matching of anode and electrolyte sintering rates was obtained, resulting in a virtually unconstrained co-sintering and a cell which do not develop any significant curvature during the whole firing process, thus demonstrating the feasibility of a straightforward anode supported half-cell production. SEM micrographs demonstrated the absence of defects in the samples possessing low sintering rate mismatch between the constituting layers, while when the sintering rate mismatch is high, severely damaged microstructures are developed.

REFERENCES

[1]K. C. Wincewicz, J. S. Cooper, "Taxonomies of SOFC material and manufacturing alternatives", J. Power Sources, 140 280–296 (2005).
[2]F. Tietz, Q. Fu, V. A. C. Haanappel, A. Mai, N. H. Menzler, S. Uhlenbruck "Materials Development for Advanced Planar Solid Oxide Fuel Cells", Int. J. Appl. Ceram. Technol., 4 [5] 436–445 (2007).
[3]E. Ivers-Tiffée, A. Weber, D. Herbstritt, "Materials and technologies for SOFCcomponents", J. European. Ceram. Soc., 21, 1805-1811 (2001).
[4]J. Kanters, U. Eisele, J. Rödel "Cosintering simulation and experimentation: case study of nanocrystalline zirconia" J. Am. Ceram. Soc., 84 [12] 2757-63 (2001).

[5]G-Q Lu. R. C. Sutterlin, T. K. Guopta, "Effect of Mismatched Sintering Kinetics on Camber in a Low-Temperature Cofired Ceramic Package", *J. Am. Ceram. Soc.,* 76 [8] 1907-14 (1993).

[6]K. R. Venkatachari, R. Raj, "Shear deformation and densification of powder compacts", *J. Am. Ceram. Soc.,* 69 [6] 499-506 (1986).

[7]R. K. Bordia, R. Raj, "Sintering behavior of ceramic films constrained by a rigid substrate", *J. Am. Ceram. Soc.,* 68 [6] 287-292 (1985).

[8]J-B. Ollagnier, O. Guillon, J. Rödel, "Viscosity of LTCC determined by discontinuous sinter forging", *Int. J. Appl. Ceram. Technol.,* 3 [6] 437-441 (2006).

[9]C. Hillman, Z. Suo, F. F. Lange, "Cracking of laminates subjected to biaxial tensile stresses", *J. Am. Ceram. Soc.,* 79 [8] 2127-33 (1996).

[10]V. M. Sglavo, P. Z. Cai, and D. J. Green, "Damage in Al_2O_3 Sintering Compacts under Very Low Tensile Stress", *J. Mat. Sci. Let.,* 18 895-900 (1999).

[11]T. Garino, "The Co-Sintering of Electrode and Electrolyte Layered Structures for SOFC Applications", *Ceram. Eng. and Sci. Proceedings,* 23 759-766 (2002).

[12]P. Z. Cai, G. L. Messing, and D. L. Green, "Constrained Densification of Alumina/Zirconia Hybrid Laminates, I: Experimental Observations of Processing Defects", *J. Am. Ceram. Soc.,* 80 1929–1939 (1997).

[13]P. Z. Cai, G. L. Messing, and D. L. Green, Constrained Densification of Alumina/Zirconia Hybrid Laminates, II: Viscous Stress Computation, *J. Am. Ceram. Soc.,* 80 1940–1948 (1997).

[14]C. Müller, D. Herbstritt, and E. Ivers-Tiffée, "Development of a multilayer anode for solid oxide fuel cells", *Solid State Ionics* 152 537–542 (2002).

[15]L. Jia, Z. Lub, J. Miaob, Z. Liu, G. Li, W. Sua, and "Effects of pre-calcined YSZ powders at different temperatures on Ni–YSZ anodes for SOFC", *J. Alloys Compd.* 414 152–157 (2006).

[16]A. Atkinson, and A. Selcuk, "Residual stress and fracture of laminated ceramic membranes", *Acta Mater.* 47 867–874 (1999).

[17]D. Montinaro, S. Modena, S. Ceschini, M. Bertoldi, T. Zandonella, A. Tomasi, and V. M. Sglavo, "Anode Supported Solid Oxide Fuel Cells with Improved Cathode/Electrolyte Interface", *Ceramic Transactions,* 179 139-147 (2006).

[18]S-H. Lee, G. L. Messing, M. Awano, "Sintering Arches for Cosintering Camber-Free SOFC Multilayers", J. Am. Ceram. Soc., 91 [2] 421-427 (2008).

[19]D. Pohle, M. Wagner, A. Roosen, "Effect of Friction on Inhomogeneous Shrinkage Behavior of Structured LTCC", J. Am. Ceram. Soc., 89 [9] 2731-2737 (2006).

[20]M. Cologna, M. Bertoldi, V. M. Sglavo, "Water-Based Tape Casting and Co-Sintering of Bi-layers for SOFC Applications", Proceedings of the 2nd International Congress on Ceramics, Verona, 29/6 - 4/7/2008, CD edition.

[21]S. Modena, S. Ceschini, D. Montinaro, M. Bertoldi, Development and Characterization of Doped Ceria Buffer Layers, proceedings of the 8th European Solid Oxide Fuel Cell Forum, 30 June – 4 July 2008, Lucerne, CH.

Table 1. Slurry composition (wt%).

	Electrolyte	Anode
Powder	69	61
Dispersant	3	4
Water	16	10
Binder	12	25

Table 2. True strain and strain rate data for the electrolyte and anode tapes.

	Ec	Ef	Ac	Af
Temperature at 1%ε [°C]	1230	1200	1240	1220
ε at 1300°C	-0.03	-0.05	-0.03	-0.04
ε at 1450°C (0h)	-0.12	-0.21	-0.12	-0.15
ε at 1450°C (4h)	-0.18	-0.25	-0.17	-0.21
$|d\varepsilon/dt|$ max [10^{-3}min^{-1}]	-0.77	-1.34	-0.79	-1.00
Temperature at $|d\varepsilon/dt|_{max}$[°C]	1420	1380	1440	1420

Fig. 1. SEM micrograph of the selected powders. NiO (a), calcined NiO (b), coarse YSZ (c), fine YSZ (d).

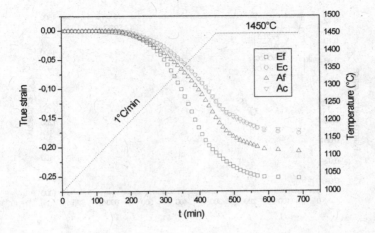

Fig. 2. True strains as a function of time for tapes Ef, Ec, Af and Ac.

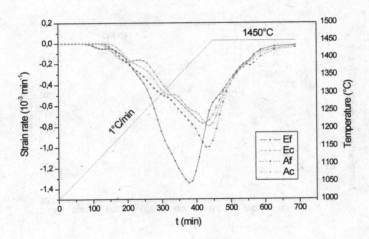

Fig. 3. Strain rate as a function of time for tapes Ef, Ec, Af and Ac.

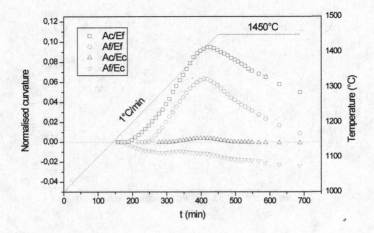

Fig. 4. Normalised curvature as a function of time for the produced half cells.

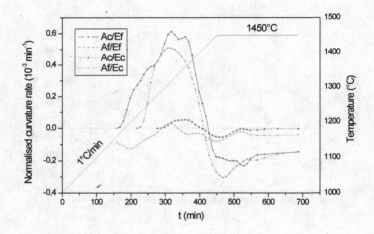

Fig. 5. Normalised curvature rate as a function of time for the produced half cells.

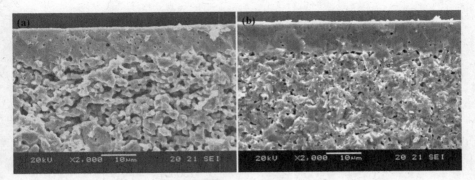

Fig. 6. Cross section of the as sintered cells. Ac/Ec (a) and Af/Ec (b).

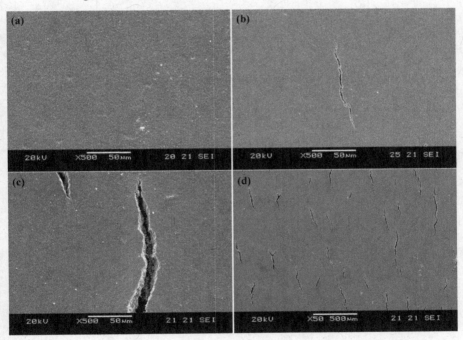

Fig. 7. Electrolyte surface of the as sintered cells. Ac/Ec (a), Af/Ef (b), Ac/Ef (c), Ac/Ef at a lower magnification (d).

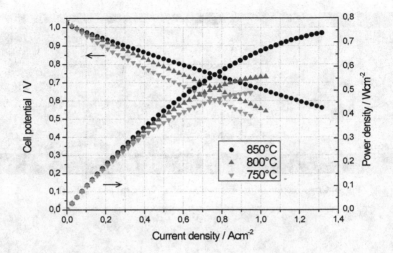

Fig. 8. Electrochemical test of sample Ac/Ec.

PHASE-BOUNDARY GROOVING AT SURFACES OF SOLID OXIDE FUEL CELL MATERIALS

Sanjit Bhowmick, Jessica L. Riesterer, Yuan Xue and C. Barry Carter

Department of Chemical, Materials and Biomolecular Engineering, University of Connecticut, Storrs, CT 06269, USA

ABSTRACT

Thermal grooving was studied on the interfaces between two ceramic phases. Specimens of two-phase materials, lanthanum strontium manganite (LSM) and cerium oxide, 50–50 vol%, were prepared by mechanical milling and sintered at 1400 °C for 24 hrs. LSM and ceria are two important components for new generation solid oxide fuel cells (SOFC), and the boundary between these two and gas phase, forming a triple phase boundary (TPB), plays a significant role in the performance of a fuel cell. For the thermal grooving study, LSM-ceria specimens were polished and heat treated at 1400 °C for three annealing times. The phase boundaries were characterized and monitored by FESEM and AFM. It has been noticed that the formation and evolution of grooving at two-phase interfaces are similar to those observed in grain boundaries of single-phase systems. The preliminary data of the groove dimensions are presented.

KEY WORDS: Phase boundary, Thermal grooving, Triple phase boundary, LSM, Ceria

INTRODUCTION

In solid oxide fuel cells (SOFC) and polymer electrolyte membrane fuel cells (PEMFC), triple-phase-boundaries (TPB) are considered as the most active sites for electrode reactions[1-6]. Generally, TPBs control the hydrogen oxidation reaction and oxygen reduction reaction at particular interfaces of porous electrode, dense electrolyte and gas. Sometimes these boundaries are considered as thick wide zones rather than one-dimension interfaces[7-9]. Understating the nature and properties of these zones is important as the TPBs control diffusion process and reactions[10].

Thermal grooving is a process driven by capillary at the regions where the grain boundary intersects the free surface. In this process, the transportation of materials may occur by surface diffusion along the interfaces, by volume diffusion through bulk or by evaporation–condensation mechanism[11-16]. Thermal grooving studies at the grain boundaries have been used to determine diffusion mechanisms and kinetics, interfacial energies and surface energies in many ceramics and metals[17-20]. The grooving at the interfaces between two phases can also be used to understand diffusion kinetics, dimension and shape of the TPB zone, high temperature stability of the interfaces, reaction products and element segregations of fuel cell materials. In this study, the specimens with LSM and ceria grains (50-50 vol%) were prepared by solid-state synthesis route. The phase boundaries intersecting the free specimen surface are representative of TPBs in SOFC. A combination of atomic-force microscopy (AFM), visible-light microscopy (VLM) and field-emission scanning electron microscopy (FESEM) was used to observe phase boundary grooving and microstructural morphologies. This paper reports, for the first time, some preliminary observations and data of the grooves along the interfaces between two phases.

EXPERIMENTAL

$La_{0.9}Sr_{0.1}MnO_3$ (LSM) powder was prepared by conventional solid-state synthesis route. Appropriate stoichiometric weight of high purity lanthanum (III) oxide (99.99%, Acros Organics, New Jersey), strontium carbonate (Johnson Matthey, Royston, UK) and Manganese carbonate (Johnson Matthey, Royston, UK) were mixed and ball milled in isopropyl alcohol for 24 hrs. The dried powder was subsequently calcined at 800 °C for 10 hrs in air. Alumina balls were used as medium for ball milling. The calcined LSM powder was hand mixed with 50 vol% cerium oxide (99.99%, Alfa Aesar, Ward Hill, MA) in mortar and pastel. The mixed powder was uniaxially pressed to make approximately 12 mm diameter and 2 mm thick pellet. The pellets were sintered at 1400 °C for 24 hrs and then diamond-polished to 0.5 μm finish. For the thermal grooving experiments, the polished samples were placed inside an aluminum crucible on a platinum foil. The heat treatments were carried out inside a box furnace at 1400 °C with holding times of 0.1, 0.5 and 1 hr, at a heating and cooling rate of 20 °C/min.

X-ray diffraction of the calcined LSM powder and sintered pellets were carried out on a Bruker-Axs D5005 automated diffractometer, over a range 2θ of 10-70° with scanning rate 2°/minute, using Cu K_α radiation, with generator voltage and current settings of 40 kV and 40 mA. A combination of VLM and FESEM (JEOL 6335F) was used to map the region of interest so that same grains and phase boundaries can be traced by AFM after each heat treatment. AFM images were obtained in contact mode using Park Systems XE-70 with decoupled XY and Z Scanners (Park Systems Corp, Santa Clara, CA). Point Probe plus silicon AFM probes (Nanosensors™, Neuchatel, Switzerland) with a tip radius <10 nm were used in imaging. The images and groove profiles were analyzed and measured using the software XEI (Park Systems Corp, Santa Clara, CA).

RESULTS AND DISCUSSION

X-ray diffraction pattern of $La_{0.9}Sr_{0.1}MnO_3$ powder is shown in fig. 1. The diffraction pattern confirms the rhombohedral phases of LSM. Fig. 2 shows SEM image sequences of annealed LSM-ceria specimen after heat treatment at 1400 °C for t = 0.1, 0.5 and 1 hr. Fig. 2b is a representative illustration of the LSM (white) and ceria (gray) grains taken from fig. 2a. XEDS (X-ray energy dispersive spectroscopy) taken on each grain (data not shown) confirms that the grains with step-like faceted morphology are LSM (marked 'A' in fig 2b) and the grains with no steps but dimple-like morphology are ceria (marked 'B' in fig 2b). In this study, we choose five phase boundaries between one large LSM grain (A1) and five ceria grains (B1–B5). The morphology of the interfaces and grains changes after heat treatment t = 0.5 hr (fig. 2c) compared to t = 0.1 hr (fig. 2a). Severe disruptions of the LSM grains, especially close to the boundaries are apparent (A2, A3 and A4). The steps on LSM grains become more pronounced with subsequent heat treatments. In some LSM grain boundaries, an additional layer appears close to the main boundary (e.g. A1-B1 and A1-B4 in fig. 2c). Note that the ceria-ceria grain boundaries remain relatively unaffected, even after annealed for 1 hr (fig. 2d). The observed morphology of the microstructure for t = 1 hr were similar to t = 0.5, except that the LSM grains were markedly more disruptive. No grain or phase boundaries movement has been noticed in the present experimental condition.

AFM deflection images of this region are shown in fig 3a-3c, and grove profiles at the boundary between A1-B3 are plotted from the topography images in fig 3d-3f. More details about the nature and shape of the grooves are revealed in the cross-section profile data. The profile shows the presence of an extra layer at the left of the boundary after t = 0.5 hr in fig. 3e, which disappears after t = 1 hr in fig 3f. Surface ridges on both sides of the phases are evident with larger heights on the LSM grains. The dihedral angles of the groves at the phase boundaries were found to be slightly asymmetric; the inclined boundaries with LSM are observed to be little higher than the ceria boundaries. The shape of the phase boundary grooves is found to be similar to those observed in grain boundaries of single-phase systems.

Phase boundary widths W are measured between the ridges and plotted as a function of annealing times t in fig. 4a. Five grain boundaries are considered for the statistical analysis in this study. Only linear boundaries are considered in the measurements; data close to the triple point boundaries are avoided. The data points are average of five to ten measurements with standard deviations, each symbol representing a particular annealing time. The data show some scatter, indicative of groove width variation for a particular phase interface. Measurements of the groove width (W) at different annealing time (t) are re-plotted on log-log scale for five boundaries in fig. 4b. The dotted lines in this figure represent the linear dimensions of the grooves with time t, as $t^{1/3}$ and $t^{1/4}$, where t is annealing time. The lines are drawn assuming that LSM and ceria have similar diffusion kinetics (diffusion data for LSM and ceria are not available). The lines are predictions for bulk diffusion ($t^{1/3}$) and surface diffusion ($t^{1/4}$)[11-12]. Note that the line $t^{1/4}$ fits the data reasonably well for the B3, B4 boundaries. For B1 and B2 boundaries, the data are lower than the prediction, whereas for B5 boundary, the data lies above. The initial groove width at 1400 °C for 0.1 hr for all five boundaries varies between 4 – 6 μm. The groove width increases for all the boundaries between annealing time 0.1 hr and 1 hr; maximum increase of 72% observed for the boundary A1-B3, lowest increase of 12% observed for A1-B1. The other three boundaries, A1-B2, A1-B4 and A1-B5 show 39%, 57% and 67% increase, respectively. From these preliminary results, one can conclude that several mechanisms may operate under present condition but the dominant mechanisms could be surface diffusion, at least for the boundaries between A1-B3, A1-B4 and A1-B5. The variation in the increasing rates also confirms the dependency of phase orientations on the groove evolution. A TEM study at the boundaries between LSM and Ceria is necessary to establish relationships between crystallographic orientation of the phases and groove dimensions. An elemental mapping across the phase boundaries before and after groove formation will provide the information about segregation of the elements near the interface. These studies are currently underway.

CONCLUSION

An understanding of mass transfer mechanisms in multi-phase systems is very important for high temperature applications. However, the predictions of the diffusion mechanism for two phases are not straightforward, requiring an understanding of diffusion kinetics of the elements. Our attention in this paper is focused on presenting preliminary experimental data on phase boundary grooving and correlating the behavior of phase interfaces with grain boundaries. The main conclusion that follows from the experimental results is that the evolution of grooving at two-phase interfaces appears similar to that observed in grain boundaries of single-phase systems. The groove widths for all boundaries increase with temperature, but the increasing rate varies depending upon orientation of the phases.

REFERENCES
[1] S. P. Jiang and S. H. Chan, A review of Anode Materials Development in Solid Oxide Fuel Cells, *J. Mater. Sci.*, **39**, 4405(2004).
[2] L. Carrette, K. A. Friedrich and U. Stimming, Fuel Cells: Principles, Types, Fuels, and Applications, *Chemphyschem*, **1**, 162(2000).
[3] S. M. Haile, Fuel Cell Materials and Components, *Acta Mater.*, **51**, 5981(2003).
[4] M. Brown, S. Primdahl and M. Mogensen, Structure/Performance Relations for Ni/Yttria-Stabilized Zirconia Anodes for Solid Oxide Fuel Cells, *J. Electrochemical Soc.*, **147**, 475(2000).
[5] S. P. Jiang, Development of Lanthanum Strontium Manganite Perovskite Cathode Materials of Solid Oxide Fuel Cells: A Review, *J. Mater. Sci.*, **43**, 6799(2008).
[6] S. J. Skinner, The Materials Science of Fuel Cells, *J. Mater. Sci.*, **36**, 1051(2001).
[7] R. O'Hayre, D. M. Barnett and F. B. Prinz, The Triple Phase Boundary - A Mathematical Model and Experimental Investigations for Fuel Cells, *J. Electrochemical Soc.*, **152**, A439(2005).
[8] V. Janardhanan, V. Heuveline and O. Deustschmann, Three-Phase Boundary Length in Solid-Oxide Fuel Cells: A Mathematical Model, *J. Power Sources*, **178**, 368(2008).
[9] K. V. Jensen, R. Wallenberg, I. Chorkendorff and M. Mogensen, *Solid State Ionics*, **160**, 27 (2003).
[10] Y. L. Liu and C Jiao, Microstructural Degradation of an Anode/Electrolyte Interface in SOFC

Studied by Transmission Electron Microscopy, *Solid State Ionics*, **176**, 435 (2005).

[11] W. W. Mullins, Theory of Thermal Grooving, *J. Appl. Phys.*, **28**, 333(1957).

[12] W. W. Mullins, The Effect of Thermal Grooving on Grain Boundary Motion, *Acta Metall.*, **6**, 414(1958).

[13] M. P. Mallamaci and C. B. Carter, Faceting of the Interface Between Al2O3 and Anorthite Glass, *Acta Mater.*, **46**, 2895(1998).

[14] W. Shin, W. S. Seo, and K. Koumoto, Grain-Boundary Grooves and Surface Diffusion in Polycrystalline Alumina Measured by Atomic Force Microscope, *J. Eur. Ceram. Soc.*, **18**, 595(1998).

[15] C. B. Carter and M. G. Norton, Ceramic Materials: Science and Engineering, Springer, New York, 2007.

[16] N. E. Munoz, S. R. Gilliss and C. B. Carter, The Monitoring of Grain-Boundary Grooves in Alumina, *Philos. Mag. Lett.*, **84**, 21(2004).

[17] N. E. Munoz, S. R. Gilliss and C. B. Carter, Remnant Grooves on Alumina Surfaces, *Surface Science*, **573**, 391(2004).

[18] D. M. Saylor and G. S. Rohrer, Measuring the Influence of Grain-Boundary Misorientation on Thermal Groove Geometry in Ceramic Polycrystals, *J. Am. Ceram. Soc.*, **82**, 1529(1999).

[19] E Saiz, R. M. Cannon and A. P. Tomsia, Energetics and Atomic Transport at Liquid Metal/Al2O3 Interfaces, *Acta Mater.*, **47**, 4209(1999).

[20] D. B. Marshall, J. R. Waldrop and P. E. D. Morgan, Thermal Grooving at the Interface Between Alumina and Monazite, *Acta Mater.*, **48**, 4471(2000).

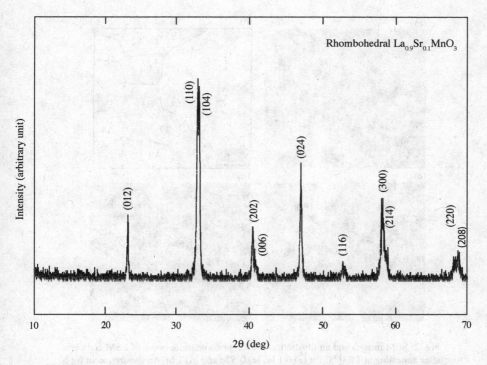

Fig. 1: X-ray diffraction of La0.9Sr0.1MnO3 powder after calcinations at 800 °C for 10 hrs.

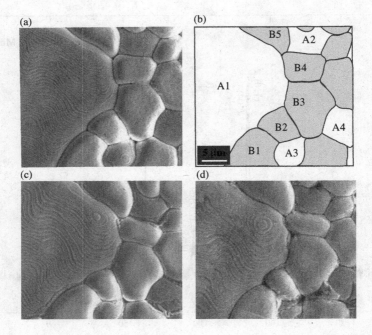

Fig. 2: SEM images and an illustration showing two phase network of LSM and ceria after annealing at 1400 °C for (a) 0.1 hr, (c) 0.5 hr and (d) 1 hr. An illustration in fig b indicates the position of the grains. Step-like faceted morphology appears on LSM grains (white in the illustration), and dimple-like morphology on ceria grains (gray in the illustration). The interfaces characterized in this study are A1-B1, A1-B2, A1-B3, A1-B4 and A1-B5.

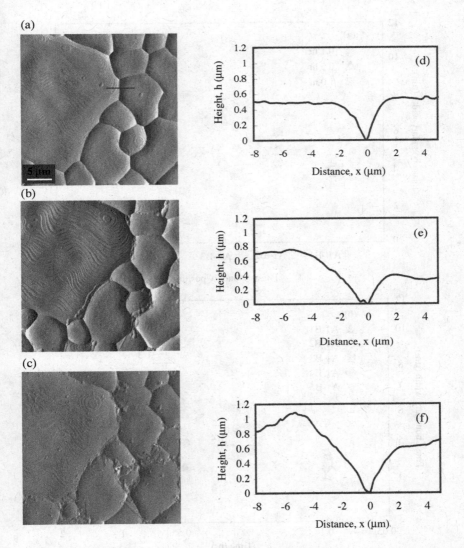

Fig. 3: AFM images of LSM-ceria grains after annealing at 1400 °C for (a) $t = 0.1$ hr, (b) 0.5 hr and (c) 1 hr; (d-f) groove profile of between the grains A1-B3.

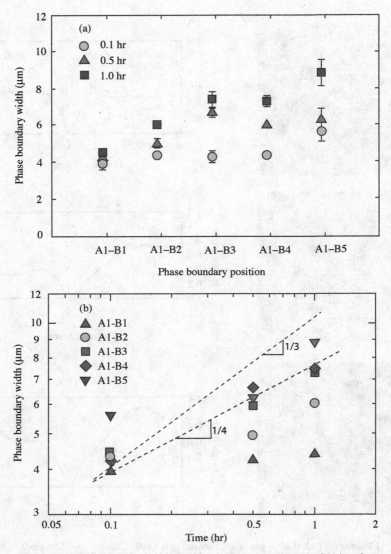

Fig. 4: (a) Plot of phase groove width W for three annealing times at five interfaces. (b) Plot of phase groove width W as a function of annealing time t. Experimental data points in (a) are means and standard deviations.

Electrodes

MIXED PROTON-OXIDE ION-ELECTRON CONDUCTING CATHODE FOR SOFCS BASED ON OXIDE PROTON CONDUCTORS

Lei Yang, Shizhong Wang, Ze Liu, Chendong Zuo, Meilin Liu

School of Materials Science and Engineering, Georgia Institute of Technology, Atlanta, GA, 30332-0245, USA
Phone: (404) 894-6114; Fax: (404) 894-9140
E-mail: meilin.liu@mse.gatech.edu

ABSTRACT

The objective of this study is to develop compatible cathode for $Ba(Zr_{0.1}Ce_{0.7}Y_{0.2})$ $O_{3-\delta}$ (BZCY) based proton conducting SOFC. A new concept cathode is proposed for proton conductor, that is, proton-oxygen ion-electron mixed conducting cathode. Introduction of proton conduction into conventional oxygen ion-electron conducting cathode greatly expand the reaction sites and thus decrease the interfacial resistance.

1. INTRODUCTION

It is of paramount importance to develop a highly conducting electrolyte to lower the operating temperature of SOFCs from the traditional 1000 °C to 600–800°C [1]. Oxide proton conductor, which exhibits high conductivity and superior chemical stability, has a potential to meet most of the requirements for ITSOFCs [2]. The recent progress in the development of high conductivity proton conductors is encouraging despite the need for further fundamental investigation and stability testing. However, the performance of the cells reported so far are still negative due to the delay of the development of promising electrodes compatible with the electrolyte and the lack of improved fabrication techniques for preparing the electrode supported thin film [3]. Ni is expected to be a high performance anode for proton conducting SOFCs since Ni - yttrium doped barium cerate is a well know mixed conductor for hydrogen permeation. However, few positive results are available on the development of cathode for proton conducting SOFCs. Pt, Ag, $Ba_xPr_{1-x}CoO_3$, and some conventional perovskite mixed ionic-electronic conducting cathodes showed rather poor performance. Hibino [4] utilized SSC and LSC cathodes for Y doped barium cerate electrolyte, which yielded overpotential of 350 and 270 mV at 100 mA cm^{-2} under 600 °C. These values are much larger than the overpotential of SSC cathodes on GDC electrolyte at same conditions.

Recently, we have developed a new composition of oxide proton conductor, $Ba(Zr_{0.1}Ce_{0.7}Y_{0.2})$ $O_{3-\delta}$ (BZCY), in the Barium-Zirconium-Cerium-Yttrium family which exhibits not only adequate proton conductivity but also sufficient chemical and thermal stability over a wide range of conditions relevant to SOFC operation [5, 6]. A proton-oxygen ion-electron conducting cathode formed by conventional $Sr_{0.5}Sm_{0.5}CoO_{3-\delta}$ (SSC) and $La_{0.6}Sr_{0.4}Co_{0.2}Fe_{0.8}O_{3-\delta}$ (LSCF) cathode and BZCY composite is investigated. The cell with this composite cathode demonstrates remarkable power density under fuel cell operation conditions. It was proposed that the unique transport properties provided by mixed conducting cathode greatly increases the number of active sites, facilitating the electrochemical reactions involving H^+, O^{2-} and e' or $h^{\bullet}$.

2. EXPERIMENTAL

The powders of BZCY and NiO at a weight ratio of 1: 1.92 (volume ratio of 50: 50) were weighted and mixed with starch as pore former and organic binder to form an anode precursor. The precursor was pressed in a stainless module to form anode substrate with flat surface, and to possess certain mechanical strength. Then the BZCY powders were added and evenly distributed onto the substrates. The coated BZCY powders layer was pressed together with the anode substrate and the bilayer was subsequently sintered at 1350°C for 6 h, and then dense BZCY7 electrolyte films were formed. The BZCY was dense and crack free as reported previously, and the thickness of the electrolyte film was about 60 μm. The morphologies of the fuel cells were examined using a scanning electron microscope (SEM, Hitachi S-800).

Cathode of SSC-BZCY and LSCF-BZCY was screen-printed on the BZCY electrolyte film and sintered at 1000°C for 2 h to form a single cell. The weight ratio of SSC (LSCF) to BZCY is about 70:30. Two silver wires were pasted on each electrode as current leads. The MEA was sealed on one side of a ceramic tube to finish the fabrication of a single cell. The cell was tested at 400-750°C with humidified hydrogen (3 vol% H_2O) as fuel and stationary air as oxidant. The fuel cell performances were measured using an EG&G Potentiostat/Galvanostat (Model 273A) interfaced with a PC through electrochemical impedance software (Model 398).

3. RESULTS AND DISCUSSIONS

3.1 Cathode reactions

While proton conductor exhibits high ionic conductivity, they are not suitable for SOFCs due to lack of compatible cathode. The electrodes performances are not sufficient for practical applications. This may be due to the possibility that triple phase boundary is limited to the electrolyte surface when oxygen ion –electron conducting cathode is applied to a proton conductor [7]. As shown in figure 1, the reaction of oxygen ion, proton and electron can only occur at the certain points on the electrolyte surface. The formed water is also difficult to flow out. Thus, the performance of single oxygen ion conducting cathode is expected to be very poor during cathode reaction. Indeed, single SSC cathode has been previously shown to produce a very large overpotential for Y doped $BaCeO_3$ electrolyte, although it is a superior cathode for GDC based SOFCs. On the contrary, a great change might happen when proton conduction is added to the conventional oxygen ion-electron conducting cathode. Figure 1 (b) shows the possible reactions of proton-oxygen ion-electron mixed conducting cathode on a proton conductor. It is easy to find that the reaction sites are greatly expanded to entire cathode. The interfacial resistance should be decreased. However, it may be difficult to get a single phase with mixed proton-oxygen ion-electron conduction required for transport of various charge carriers. Especially, a good proton and electron conducting phase has not been reported so far. Thus, from the ideal of new concept cathode, we designed a proton-oxygen ion-electron conducting composite phase through addition of proton conductor to the conventional oxygen ion conductor. The contact areas of proton and oxygen ion-electron particles could be utilized as the reaction sites. BZCY-SSC and BZCY-LSCF cathode were applied in our research. This composite is totally different from GDC-SSC cathode, in which GDC is just used to enhance the oxygen ion conductivity of the cathode. The extremely low interfacial resistance was observed comparing with the single oxygen ion conducting cathode. Moreover, to the best of our knowledge, the as-prepared cell demonstrated the highest peak power density in proton conducting SOFCs.

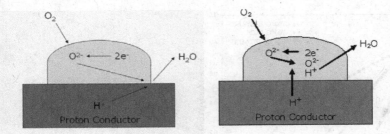

Figure 1 Schematic illustration of cathodic reaction mechanism of (a) oxide ion-electron conducting (b) proton-oxide ion-electron conducting cathodes on a proton conductor.

3.2 Power density

Shown in Figure 2 are the current-voltage characteristics and corresponding power densities for the cells with BZCY-SSC, BZCY-LSCF, GDC-SSC and LSCF cathodes at 700 and 600 °C. Each datum point was recorded about 2h after cell reached a steady state. The open circuit voltage (OCV) is about 1.04 V at 600°C, and 1.08 V at 500°C, indicating that the electrolyte thin film prepared by the dry processing process is dense enough. The electronic conductivity of the electrolyte is negligible under the fuel cell conditions. Obviously, BZCY-SSC cathode shows much higher power output than GDC-SSC cathode. The maximum power densities of BZCY-SSC cell are 0.72 and 0.45 W. cm^{-2} at 700 and 600°C, respectively. On the contrary, GDC-SSC cell merely produced peak power density of 380 and 175 mW/cm^2 at 700 and 600 °C. BZCY-LSCF also exhibited better performance than single LSCF cathode, although the peak power density is slightly lower than BZCY-SSC cathode. The performance discrepancies indicate that addition of proton conductivity effectively accelerates oxygen reduction reaction, thereby decreasing the cathodic interfacial resistances. Mixed proton-oxide ion-electron conducting cathode is suitable for SOFCs based on oxide proton conductors.

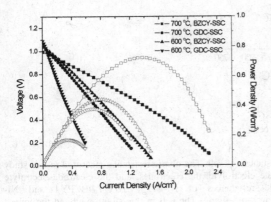

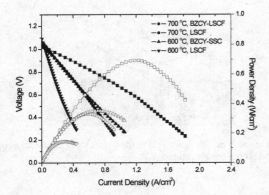

Fig. 2 V-I performance curve for BZCY based anode-supported cells with BZCY-SSC (a) and BZCY-LSCF (b) cathodes measured in humidified (3% H_2O) H_2 at 700 and 600 °C

3.3 Ohmic and interfacial resistances

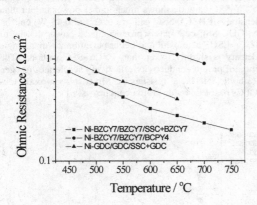

Fig.3 The ohmic resistances of various cells

Shown in Fig.3 are the ohmic resistances of cell. The ohmic resistance measured in this study should include the ohmic resistance of the electrolyte, the electrodes, and the electrode/electrolyte interface. For comparison, the ohmic resistances of Ni-BZCY7/BZCY7/BCPY4 and Ni-GDC/GDC/SSC-GDC [8] are shown in the same figure. The cell in this study exhibited the lowest ohmic resistance among the three cells. It is even lower than the cell based on 15 μm GDC electrolyte.

It is clear that the high ohmic resistance of the cell with a BCPY4 cathode is due to the ohmic resistance of the cathode and, probably, the cathode/electrolyte interface. The ohmic resistance calculated from the conductivity of BZCY reported previously is close to the experimental value shown in Fig. 3, suggesting that the ohmic resistance is mainly determined by the ohmic resistance of the electrolyte. For example, the calculated ohmic resistance at 500 ^oC is about 0.7 $\Omega.cm^2$, which is very close to the 0.6 $\Omega.cm^2$ shown in Fig.3. It is needed to be emphasized that the ohmic resistance of protonic conductor under fuel cell conditions could be much lower than that measured in H_2 saturated with water at room temperature due to the possible high concentration of water inside the anode/electrolyte/cathode assembly. This is especially true for the cells with high performance. This is one of the advantages of applying protonic conductors as the electrolyte for fuel cells over conventional oxygen ion conductors. It is expected that the ohmic resistance would be further minimized dramatically by using our modified dry pressing process, which could make a film as thin as 15 μm [8].

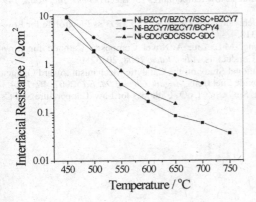

Fig.4 The interfacial polarization resistances of various cells.

Shown in Fig.4 are the interfacial polarization resistances of various cells. The interfacial polarization resistance of Ni-BZCY/BZCY/SSC-BZCY is the lowest among the three cells at low temperatures, demonstrating the high activity for oxygen reduction of SSC-BZCY cathode. However, with the decrease in temperature, the interfacial resistance increases sharply and reaches the same as that applying a BCPY4 cathode. It is clear that the electrodes of Ni-BZCY/BZCY/SSC-BZCY need to be further optimized. At this time, it is still not clear which electrode is responsible for the high interfacial resistance shown in Fig.4. A careful designed three-electrode measurement is essential to separate the interfacial resistance of each electrode from the overall polarization resistance. Comparing the ohmic resistance in Fig.3 with the interfacial polarization resistance shown in Fig.4, it can be seen that the cell performance is determined by the ohmic resistance at high temperatures, and by interfacial polarization resistances at low temperatures. To develop high performance low temperature cells for vehicle application, it is essential to design and optimize new electrodes.

The present study systematically investigated the mixed proton-oxygen ion-electron conducting composite cathode for proton conducting SOFC. Introduction of proton conductivity in the cathode greatly extend the reactions sites and thus facilitates oxygen reduction reaction. With these novel composite cathodes, the BZCY based cell produced highest power density for any SOFCs based on oxide proton conductors ever reported under the testing conditions. However, the long term stability of BZCY-SSC and BZCY-LSCF composite cathodes should be investigated before the practical application.

REFERENCES

[1] N. Q. Minh, Ceramic fuel cells, *J. Am. Ceram. Soc.* 1993, 76, 563.

[2] K.D. Kreuer , S.J. Paddison , E. Spohr, M. Schuster, Transport in Proton Conductors for Fuel-Cell Applications: Simulations, Elementary Reactions, and Phenomenology, *Chem. Rev.* 2004, 104, 4637.

[3] M. Koyama, C. Wen, K. Yamada. The mechanism of porous Sm0.5Sr0.5CoO3 cathodes used in solid oxide fuel cells, *J. Electrochem. Soc.* 2000, 147, 87

[4] T. Hibino, A. Hashimoto, M. Suzuki, et al. A solid oxide fuel cell using Y-doped BaCeO3 with Pd-loaded FeO anode and Ba0.5Pr0.5CoO3 cathode at low temperatures , *J. Electrochem. Soc.* 2002, 149, A1503.

[5] C. Zuo, S. Zha, M. Liu, M. Hatano, M. Uchiyama, Ba(Zr0.1Ce0.7Y0.2)O3 as Electrolyte for Low-temperature SOFCs, *Adv. Mater.* 2006, 18, 3318.

[6] L. Yang, CD Zuo, SZ Wang, Z. Cheng. M.L. Liu, A Novel Composite Cathode for Low-Temperature SOFCs Based on Oxide Proton Conductors, *Adv. Mater.* 2008, 20, 3280.

[7] Y. M. Choi, M. C. Lin, M. Liu, Computational Study on the Catalytic Mechanism toward Oxygen Reduction on $La_{0.5}Sr_{0.5}MnO_3(110)$ in Solid Oxide Fuel Cells, *Angew. Chem., Int. Ed.*, 2007, 46, 7214.

[8] C.R. Xia, W. Rauch, F.L. Chen, M.L. Liu. $Sm_{0.5}$ $Sr_{0.5}$ CoO_3 Cathodes for Low-Temperature SOFCs, *Solid State Ionics* 2002, 149, 11.

PERMEATION AND STABILITY INVESTIGATION OF $Ba_{0.5}Sr_{0.5}Co_{0.8}Fe_{0.2}O_{3-\delta}$ MEMBRANES FOR OXY-FUEL PROCESSES

A. Ellett, D. Schlehuber, L. Singheiser and T. Markus
Forschungszentrum Jülich GmbH
Jülich, D-52425 Germany

ABSTRACT

Dense $Ba_{0.5}Sr_{0.5}Co_{0.8}Fe_{0.2}O_{3-\delta}$ disk membranes for high-purity oxygen separation in the framework of the oxy-fuel process were prepared by uniaxial pressing. The oxygen permeation properties of these membranes were investigated by mass spectrometry in the temperature range of between 700°C and 950°C. Air was fed to the membrane and argon was used as sweep gas. The oxygen permeation flux at 800°C through a 1.5 mm membrane was 1.4 $ml.min^{-1}.cm^{-2}$. Annealing experiments showed that $Ba_{0.5}Sr_{0.5}Co_{0.8}Fe_{0.2}O_{3-\delta}$ is very sensitive to CO_2- and water vapour-containing atmospheres and also reacts with chromium.

INTRODUCTION

Gas separation membranes are considered an efficient technology for future generation zero CO_2 emission power plants. The oxy-fuel process consists in the burning of coal in an oxygen-rich atmosphere to produce a flue gas highly concentrated in CO_2. Membranes will be used to separate oxygen from air in order to produce this combustion gas highly concentrated in oxygen. One of the most promising materials for application as ceramic membranes for high-purity oxygen separation is $Ba_{0.5}Sr_{0.5}Co_{0.8}Fe_{0.2}O_{3-\delta}$, a mixed ionic-electronic conducting material (MIEC) with the perovskite structure, selected for its high oxygen permeation fluxes (up to 3 ml $cm^{-2} min^{-1}$)[1,2].

In the oxy-fuel process, the flue gas will be recycled to sweep the permeate side of the membrane, exposing it to high concentrations of CO_2 and water. Therefore, the $Ba_{0.5}Sr_{0.5}Co_{0.8}Fe_{0.2}O_{3-\delta}$ membrane, operating at high temperatures, has to be stable in the gaseous atmosphere of the flue gas. The stability and the permeation of $Ba_{0.5}Sr_{0.5}Co_{0.8}Fe_{0.2}O_{3-\delta}$ are both crucial factors in the selection of this material for membrane technology in the oxy-fuel process.

In this work, the oxygen permeation of $Ba_{0.5}Sr_{0.5}Co_{0.8}Fe_{0.2}O_{3-\delta}$-based membranes was investigated as a function of temperature. Annealing experiments were performed to ensure that these materials will withstand the high temperature conditions and gases present in coal power plant emissions. Samples were exposed to CO_2, water and O_2 to check the stability of $Ba_{0.5}Sr_{0.5}Co_{0.8}Fe_{0.2}O_{3-\delta}$ in an atmosphere close to operational conditions. The degradation of the microstructure of the $Ba_{0.5}Sr_{0.5}Co_{0.8}Fe_{0.2}O_{3-\delta}$ pellets was investigated using Scanning Electron Microscopy (SEM), Electron Dispersive Spectroscopy (EDS) and X-Ray Diffraction (XRD).

OXYGEN PERMEATION

Experimental

$Ba_{0.5}Sr_{0.5}Co_{0.8}Fe_{0.2}O_{3-\delta}$ (BSCF) membranes of 15 mm diameter were uniaxialy pressed at about 100 MPa from single phased powder purchased from Treibacher Industrie AG, Austria. The green bodies were sintered at 1000°C for 10h with a heating and cooling rate of 5 K min^{-1}. The membranes were polished with successive grades of silicon carbide grinding paper to ensure flat reproducible surfaces. The permeation set-up is schematised in Figure 1. The membrane disk of 15 mm in diameter was sealed between two quartz glass tubes using two gold gaskets 1 mm thick. To ensure gas tightness, the assembly was compressed by a spring load system and heated to a temperature close to the melting point of gold in a vertical tube furnace. Synthetic air was fed to the feed side of the membrane and argon (99.5% purity) to the permeate side.

The gas flow rates were controlled by mass-flow meters (Brooks, The Netherlands). The flow rates were 100 ml min^{-1} and 50 ml min^{-1} on the feed side and on the sweep side of the membrane respectively.

The oxygen permeation rate was measured from 700°C to 950°C using a quadrupole mass spectrometer (Omnistar, Pfeifer, Germany). The leakage was determined by measuring the amount of N_2 in the permeate stream and deducted from the measured O_2 concentration. The resulting oxygen permeation flux, Jo_2 (ml cm^{-2} min^{-1}) was calculated according to equation (1):

$$j_{O_2} = \frac{FC}{S}\alpha \qquad (1)$$

where C is the permeating oxygen concentration in the argon flow (ppm) after correction for leakage; F, the argon flow rate (ml min^{-1}); S, the membrane effective surface area (cm^2) and α, a coefficient of normalization for standard temperature and pressure.

Results

At high temperatures, oxygen vacancies become available for oxygen transport through perovskite-type membranes. The temperature dependence of the permeation fluxes for the range 700°C to 950°C for a membrane of 1 mm thick is shown in Figure 2. The permeation flux increases with increasing temperature. Moreover, a linear increase is observed between 800°C and 950°C. This trend corresponds to Wagner's equation (2) that describes the oxygen flux when bulk diffusion controls oxygen permeation.

$$j_{O_2} = -\frac{RT}{16F^2L}\int_{\ln pO_2}^{\ln pO_2}\frac{\sigma_i\sigma_e}{\sigma_i+\sigma_e}d\ln(p_{O_2}) \qquad (2)$$

where σ_i and σ_e are the ionic and electronic conductivities respectively; F, the Faraday constant; R, the gas constant; T, the temperature; L, the membrane thickness and pO_2 and pO_2'', the partial pressure on the oxygen-rich and oxygen-lean side respectively.

Table I. Oxygen permeation flux (JO_2 in ml min^{-1} cm^{-2}) of BSCF membranes as a function of temperature.

Temperature (°C)	JO_2 (ml.min^{-1}.cm^{-2})
950	2,2
900	2,0
850	1,7
800	1,4
750	1,0
700	0,64

The oxygen permeation flux of BSCF reached a maximum value of 2.2 ml min^{-1} cm^{-2} at 950°C as seen in Table I. However, Bredesen and Sogge[3] reported that membranes need to exhibit oxygen permeation fluxes of at least 10 ml min^{-1} cm^{-2} in order to be competitive with the traditional high-purity oxygen separation processes.

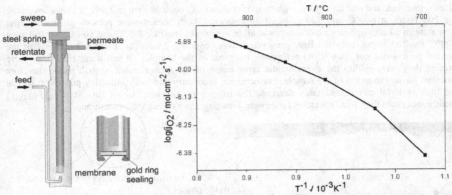

Figure 1. Membrane recipient.

Figure 2. Temperature dependence of oxygen permeation fluxes through BSCF membranes.

STABILITY INVESTIGATION

Experimental

The oxy-fuel process consists in the burning of coal in an oxygen rich atmosphere to produce a flue gas highly concentrated in CO_2. This will allow the flue gas to be composed of relatively clean exhaust gases, steam and CO_2 (80 to 85%). In this process, the flue gas will be used to sweep the permeate side of the membrane in order to collect the oxygen ions adsorbed on its surface. Therefore, the $Ba_{0.5}Sr_{0.5}Fe_{0.2}O_{3-\delta}$ membrane has to be stable in the gaseous atmosphere of the flue gas. The atmosphere that is believed to be the closest to the flue gas composition is about 89 mol% CO_2 on a dry basis, or 66 mol% CO_2 and 26 mol% water vapour, on a humid basis. Moreover, it was estimated that the permeate gas stream would consist of about 25% of permeating O_2 and 75% of flue gas.

The BSCF powder was shaped into pellets by applying around 125 MPa in a uniaxial cylindrical press. The samples weighed around 0.5 g and were 8 mm in diameter. The green-bodies were sintered at 1100°C for 5 hours. The experimental set-up is composed of a quartz glass tube placed in a three-zone horizontal tube furnace. The pellets were placed in a quartz glass sample holder and annealed at 600°C, 700°C and 800°C in the atmospheres listed below:

- Air for 200 hours and 500 hours
- Air +10 mol% CO_2 for 200 hours
- 67 mol% CO_2, 25 mol% O_2 and 8 mol% N_2 for 200 hours and 500 hours
- 50 mol% CO_2, 25 mol% O_2, 20 mol% water vapour and 5 mol% N_2 for 200 hours and 500 hours

Furthermore, given that the high temperature components in coal-fired power plants are made of chromium steel, it is of considerable importance to determine whether an interaction will occur between the construction materials and the BSCF membrane. Thus, a long term annealing experiment was carried out in which BSCF pellets were placed next to a Cr_2O_3 disk at a distance of 1 cm.

Results

After the sintering step in air, cobalt oxide (CoO) inclusions were observed in the bulk $Ba_{0.5}Sr_{0.5}Co_{0.8}Fe_{0.2}O_{3-\delta}$ pellets, as seen in Figure 3. These inclusions were composed of two different phases, one dark and one light grey, with similar contents of cobalt as seen in Table II. After annealing for 200 hours at 600°C and 700°C in all atmospheres, both phases were present. However, after annealing in all atmospheres for 500 hours at all temperatures (600°C, 700°C and 800°C) as well as at 800°C for 200 hours, only the light grey phase was observed. Moreover, the XRD analysis of the ground pellets did not show CoO reflexions for most of the samples. When these reflexions were present they were within the experimental error range of the measurement. Cobalt oxide therefore amounts for less than 1% of the sample. Furthermore, secondary non-ion-conducting phases present in the bulk material can considerably decrease the oxygen permeation flux of $Ba_{0.5}Sr_{0.5}Co_{0.8}Fe_{0.2}O_{3-\delta}$ membranes since they form obstacles for oxygen ions migrating through the membrane.

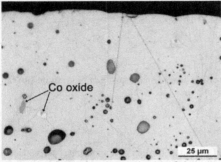

Figure 3. Light microscopy picture of the cross section of a BSCF pellet after the sintering step.

Table II Chemical composition of the two cobalt oxide phases.

Element	Phase	
	Light grey	Dark grey
Co (atomic%)	51.0	49
O (atomic%)	56.0	44.0

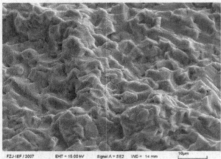

Figure 4. SEM micrograph of the $(Ba_xSr_{1-x})CO_3$ layer on the surface of a BSCF pellet annealed at 800°C for 200h in air +10% CO_2.

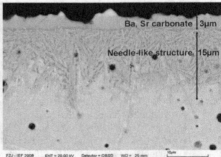

Figure 5. SEM micrograph of the cross section of a BSCF pellet annealed at 800°C for 200h in in air +10% CO_2.

After annealing in CO_2–rich atmospheres, a carbonate layer, which can prevent O_2-adsorption on the surface of oxygen permeation membranes, had built up on the surface of the $Ba_{0.5}Sr_{0.5}Co_{0.8}Fe_{0.2}O_{3-\delta}$ pellets, as seen in Figures 4, 5, 7 and 8. EDS and XRD analyses show that this layer is a mixed barium and strontium carbonate with the chemical composition $(Ba_xSr_{1-x})CO_3$. A needle-like structure layer is also formed under the carbonate layer, as seen in Figures 5 to 7. The thicknesses of both these layers were found to increase with increasing CO_2-concentration as well as annealing time and temperature (Figures 5, 7 and 8). $Ba_{0.5}Sr_{0.5}Co_{0.8}Fe_{0.2}O_{3-\delta}$ demonstrates a low stability to CO_2 since carbonates were formed on the surface of the pellets even after exposure times of only 200 hours to atmospheres with a relatively low concentration of CO_2 (Figure 4.).

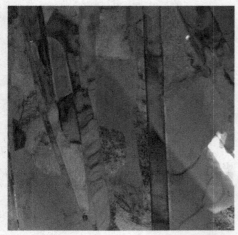

Table III. Chemical composition in mass% of a needle.

Element	Mass%
Ba	37.85
Sr	5.32
Co	50.33
Fe	6.50

Table IV. Chemical composition in mass% of the matrix material.

Element	Mass%
Ba	40.27
Sr	27.40
Co	22.04
Fe	10.29

500 nm

Figure 6. TEM micrograph of the needle-like structure.

As each individual needle in the needle-like structure layer has a maximum width of around $1\mu m$, a TEM investigation of the microstructure was conducted in order to determine the composition of this layer more precisely (Figure 6.). The chemical compositions of the needles (Table III) as well as the matrix (Table IV) were determined by EDS with a 5% error. The needles appear to be oxides with the composition $Ba_{0.38}Sr_{0.05}Co_{0.5}Fe_{0.07}O_x$ and the matrix a perovskite with the composition $Ba_{0.50}Sr_{0.53}Co_{0.64}Fe_{0.31}O_x$ which is very similar to the original $Ba_{0.5}Sr_{0.5}Co_{0.8}Fe_{0.2}O_{3-\delta}$ material.

After annealing the $Ba_{0.5}Sr_{0.5}Co_{0.8}Fe_{0.2}O_{3-\delta}$ samples in CO_2- and H_2O-containing atmospheres, a layer, which is composed of barium and strontium silicate, was formed on top of the $(Ba_xSr_{1-x})CO_3$ layer, as seen in Figure 8. After the carbonate layer started forming, a reaction with silica (SiO_2) from the quartz glass sample holder occurred as SiO_2 becomes volatile at high temperatures and high partial pressure of water vapour.

Moreover, water vapour seems to exacerbate the formation of the carbonate and needle-like structure layers as their thicknesses are higher when annealing at 800°C in an atmosphere composed of 50 mol% CO_2, 25 mol% O_2, 20 mol% water vapour and 5 mol% N_2 for 200 hours than in an atmosphere composed of 67 mol% CO_2, 25 mol% O_2 and 8 mol% N_2 for 500 hours. Even though the concentration of CO_2 decreased by about 25 percent, the thickness of the carbonate layer doubled, as seen in Figures 7(A) and 8(A). The thickness of the needle-like structure layer increased by around 90% (Figures 7(A) and 8(A)).

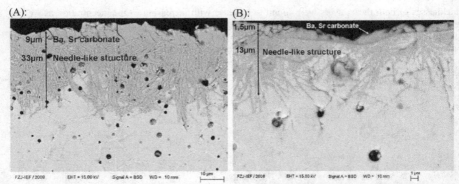

Figure 7. SEM micrographs of the cross section of BSCF pellets annealed in 67 % CO_2, 25 % O_2 and 8% N_2 for 500h: (A) at 800°C and (B) at 700°C.

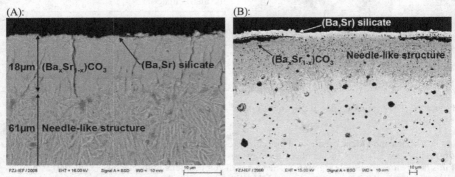

Figure 8. SEM micrographs of the cross section of BSCF pellets annealed at 800°C in 50% CO_2, 25% O_2, 20% water vapour, 5% N_2: (A) for 200h, (B) for 500h.

An investigation of the microstructure of a $Ba_{0.5}Sr_{0.5}Co_{0.8}Fe_{0.2}O_{3-\delta}$ pellet annealed at 800°C for 1000 hours in air next to a Cr_2O_3 pressed disk was carried out. EDS analysis of the surface of the pellet showed the presence of a layer which appears to be a barium and chromium oxide (Figures 9 and 10). XRD analysis determined this layer to be $BaCrO_4$. The fact that $Ba_{0.5}Sr_{0.5}Co_{0.8}Fe_{0.2}O_{3-\delta}$ reacts with chromium makes it a material incompatible for use in power plants, at least not without a protective layer.

 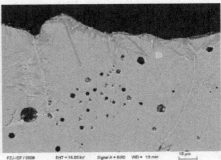

Figure 9. SEM micrograph of the surface of a BSCF pellet annealed at 800°C for 1000h in air next to a Cr_2O_3 pressed disk.

Figure 10. SEM micrograph of the cross section of a BSCF pellet annealed at 800°C for 1000h in air next to a Cr_2O_3 pressed disk.

CONCLUSION

The stability and the permeation of $Ba_{0.5}Sr_{0.5}Co_{0.8}Fe_{0.2}O_{3-\delta}$ membranes are both crucial factors in the selection of this material for membrane technology in the framework of the oxy-fuel combustion process. $Ba_{0.5}Sr_{0.5}Co_{0.8}Fe_{0.2}O_{3-\delta}$ shows very high oxygen permeation fluxes, above those of the usual perovskite membrane materials. However, the performances of these membranes are not yet sufficient to be applied in the framework of the oxyfuel process.

Moreover, $Ba_{0.5}Sr_{0.5}Co_{0.8}Fe_{0.2}O_{3-\delta}$ is very sensitive to CO_2- and water vapour-containing atmospheres. It also reacts with chromium which is a serious issue for the use of $Ba_{0.5}Sr_{0.5}Co_{0.8}Fe_{0.2}O_{3-\delta}$ membranes in coal-fired power plants. A protective layer on the surface of the membranes is therefore a possible solution to avoid the formation of these layers that would otherwise considerably decrease the oxygen permeation fluxes. The formation of cobalt oxide in the bulk of membranes $Ba_{0.5}Sr_{0.5}Co_{0.8}Fe_{0.2}O_{3-\delta}$ could also be a drawback for their use for high-purity oxygen separation.

REFERENCES

[1]H. Wang et al., Oxygen permeation study in a tubular $Ba_{0.5}Sr_{0.5}Co_{0.8}Fe_{0.2}O_{3-\delta}$ oxygen permeable membrane", Journal of Membrane Science, 210, 259-271 (2002).

[2]J.F. Vente et al., Performance of functional perovskite membranes for oxygen production, ECN-RX-05-201 (Energy efficiency in industry – Separation technology report), November 2005.

[3]cited in L. Tan et al., Influence of powder synthesis methods on microstructure and oxygen permeation performance of $Ba_{0.5}Sr_{0.5}Co_{0.8}Fe_{0.2}O_{3-\delta}$ perovskite-type membranes, Journal of Membrane Science, 212, 157-165 (2003).

NUMERICAL CONTINUUM MODELING AND SIMULATION OF MIXED-CONDUCTING THIN FILM AND PATTERNED ELECTRODES

Matthew E. Lynch, David S. Mebane, Meilin Liu
Center for Innovative Fuel Cell and Battery Technology
School of Materials Science and Engineering
Georgia Institute of Technology
Atlanta, GA, USA

ABSTRACT

The charge and mass transport processes involved in oxygen reduction reactions on a porous mixed-conducting SOFC cathode are complex and difficult to study, making analytical design of porous electrodes difficult. The use of patterned or thin-film electrode configurations is effective to isolate or separate reaction sites or pathways, making it possible to correlate electrochemical performance with electrode geometry, reaction path, or catalytically active sites. We report our recent progress in modeling of thin-film and patterned mixed-conducting electrodes by numerical solution of the continuum electrochemical and transport constitutive equations. This type of modeling allows the effects of multiple processes to be considered, reflecting the complex interdependence of underlying mechanisms. Specifically, we focus on development of multidimensional continuum models for evaluation of triple phase boundary and surface kinetics, sheet resistance effects, electrochemical impedance spectroscopy, and informed design of experiments.

INTRODUCTION

The processes that occur in the porous electrodes of solid oxide fuel cells (SOFCs) are extremely complicated and difficult to investigate and/or control. The complication arises because electrochemical reactions occur at the triple phase boundaries (TPBs) as well as on the surface of mixed ionic-electronic conducting (MIEC) electrode materials that make up the electrodes. As a consequence, the details of these processes often elude investigation using standard electrochemical experimental techniques performed on porous electrodes.

Carefully designed test cells with thin films of MIEC electrodes, with or without patterns that expose varying TPB length, offer the possibility of isolating and studying key electrochemical and transport phenomena in a geometrically simplified setting. Recently, we developed a modeling framework for the simulation of electrochemical and transport processes in thin films of MIEC materials for SOFCs.[1-4] This approach allows prediction of electrochemical response of an experimental cell from the chemical and electrical conditions imposed on it. The model is based upon a rigorous foundation in thermodynamics and kinetics from which the fundamental constitutive transport and electrochemical rate equations are derived. Two of the key rate equations can be expressed as follows.

$$r_{ads} = k_{ads}^{0} \left[\frac{P_{O_2}^{1/2}}{P_{O_{2,0}}^{1/2}} \frac{(1-\theta)}{(1-\theta_0)} \exp\left(-\frac{\alpha_{ads}F}{RT}\Delta\chi_s\right) - \frac{c_h}{c_{h,0}} \frac{\theta}{\theta_0} \exp\left(\frac{(1-\alpha_{ads})F}{RT}\Delta\chi_s\right) \right] \qquad (1)$$

$$r_{inc} = k_{inc}^{0} \left[\frac{c_v^{m}}{c_{v,0}^{m}} \frac{\theta}{\theta_0} \exp\left(\frac{\alpha_{inc}F}{RT}\Delta\chi_s\right) - \frac{c_h}{c_{h,0}} \frac{(1-\theta)}{(1-\theta_0)} \exp\left(-\frac{(1-\alpha_{inc})F}{RT}\Delta\chi_s\right) \right] \qquad (2)$$

Here, k_{ads}^{0} and k_{inc}^{0} are the equilibrium rate constants of the oxygen adsorption and incorporation reactions, respectively. The overpotential across the interface is given by $\Delta\chi$, with the subscript s denoting the MIEC/air interface. The variable θ is the fraction of filled surface sites and θ_0 is its equilibrium value. Similarly, c_v^m, c_v^e, and c_h are the surface/interfacial concentrations of vacancies and holes, while $c_{v,0}^m$, $c_{v,0}^e$, and $c_{h,0}$ are their equilibrium values (m superscript for MIEC, e for electrolyte). The partial pressure of oxygen is P_{O2} with equilibrium value $P_{O2,0}$. Finally, R is the universal gas constant, T is the temperature, F is Faraday's constant, and α is a transfer coefficient. Both of these equations assume low oxygen vacancy concentration, e.g. for a poor MIEC material such as $La_{1-x}Sr_xMnO_{3\pm x}$ (LSM), and thus only small deviations of oxygen stoichiometry from the equilibrium value. For an MIEC with larger vacancy concentration, these equations must be modified to reflect the concentration of lattice oxygen as well. Note that the equilibrium parameters we used are from the best available measurements from the literature (e.g. $c_{h,0}$).[3, 4] We postulated the parameters when experimental data was not available, e.g. for θ_0, and future work will focus on exactly determining their equilibrium value through targeted experiments evaluated using these models.

The main application of this modeling effort to date is onto cross sections of patterned electrodes (Figure 1). In this contribution, we highlight progress in several areas relevant to these models, including the simulation of TPB reaction rates, surface transport, local electrical and chemical distributions, and alternating current (AC) signal response.

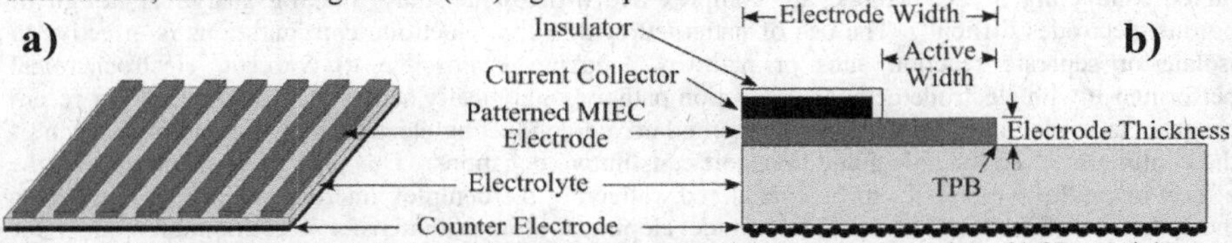

Figure 1. a) Schematic of MIEC electrodes patterned onto a solid oxide electrolyte, b) 2D partial cross section of a patterned electrode. This cross section can serve as the model domain for 2D simulations.

TRIPLE PHASE BOUNDARY AND SURFACE TRANSPORT

The TPB reaction is one of the most important reactions occurring in fuel cell cathodes and is of particular importance in SOFCs when LSM serves as the catalyst phase. It is also of particular importance in polymer electrolyte membrane (PEM) fuel cells where the catalyst is platinum. The reaction rate along the TPB contributes greatly to the fuel cell performance. We developed a numerical method to simulate TPB reaction rate[4] that works together with our previous bulk simulation[3] in order to study these reactions.

The nature of TPB reaction rate simulation is inherently multidimensional. In general, the site of the TPB reaction is a line (1D, see Figure 1a) whereas the reaction rate depends upon the chemical and electrical properties of 3D materials (MIEC, electrolyte) and their surfaces (2D). Given the right geometry, e.g. patterned electrodes, the TPB process can be simplified to a point (0D) and the MIEC and electrolyte volumes can be simplified to areas (2D) while their surfaces simplify to 1D. This simplification is illustrated by the partial cross section of a patterned electrode shown in Figure 1b, which is a representative domain for calculations. Our approach conforms to the requirements of the simplified geometry and simulates the various processes in their respective dimensions, coupling them together as boundary conditions and solving simultaneously with an in-house finite volume numerical method.[3, 4]

In Kröger-Vink notation, the oxygen reduction reaction occurring at the TPB between LSM, electrolyte, and air can be described as follows

$$O'_{ads} + V_{O,e}^{\cdot\cdot} \rightarrow O_{O,e}^x + h_m^{\cdot} + s \tag{3}$$

where O'_{ads} is a singly charged chemisorbed oxygen species, $V_{O,e}^{..}$ denotes an oxygen vacancy in the electrolyte, $O_{O,e}^x$ is an oxygen ion incorporated into the electrolyte, $h_m^.$ is an electron hole in the MIEC, and s is a surface adsorption site.

To determine the role of surface transport, we numerically solved the drift-diffusion equation for oxygen species on 1D MIEC surface coupled with generation and depletion of adsorbed oxygen due to surface adsorption and incorporation.

$$\frac{\partial c_{O^-}}{\partial t} = -\nabla \cdot \left(-D_{O^-} \nabla c_{O^-} - z_{O^-} Fu_{O^-} c_{O^-} \nabla \phi \right) + \left(r_{ads} - r_{inc} \right) \tag{4}$$

In this equation, $c_{O^-} = \Gamma\theta$ is the surface concentration of adsorbed oxygen species, Γ is the concentration of surface sites, D_{O^-} is the surface diffusion coefficient, z_{O^-} is the formal charge of the adsorbed oxygen, u_{O^-} is the mobility of the adsorbates, and ϕ is the surface electrical potential. Generation and depletion of adsorbed oxygen is determined by the rate of oxygen adsorption onto the surface, r_{ads}, and the rate of oxygen incorporation into the MIEC, r_{inc}, respectively. These values are determined by coupling with the numerical simulation of the bulk processes in the MIEC and electrolyte[3] and solved simultaneously with the 2D bulk. The simulations were implemented using in-house C++ and Matlab code.

Finally, we derived an equation for the TPB reaction rate, r_{TPB}, as a point boundary condition for the surface transport simulation.

$$r_{TPB} = k_{TPB}^0 \left[\frac{\theta}{\theta_0} \cdot \exp\left(\frac{F\Delta\chi_{es}}{2RT} \right) - \frac{c_h}{c_h^0} \frac{(1-\theta)}{(1-\theta_0)} \cdot \exp\left(-\frac{F\Delta\chi_{es}}{2RT} \right) \right] \tag{5}$$

In this equation, k_{TPB}^0 is the equilibrium rate constant and $\Delta\chi_{es}$ is the overpotential (also determined from the bulk simulation).

The development of this numerical technique allows for the bulk and TPB current paths of thin film patterned electrodes to be simulated simultaneously (required due to their interdependence), but separated and investigated independently upon completion of the simulation. The current-voltage (IV) curve in Figure 2 shows the dominance of the LSM TPB pathway at low cathodic bias but reflects the increasing importance of the bulk pathway and finally its dominance as cathodic bias increases. This behavior is in qualitative agreement with experimental evidence showing bulk activation of LSM at large cathodic bias.[5, 6] Selected numerical parameters used to generate the data in Figure 2 are shown in Table I.

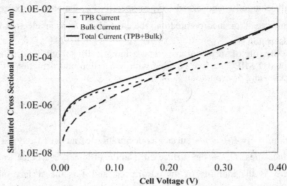

Figure 2. Current-voltage plot of a thin film patterned LSM electrode. The TPB and bulk current pathways are plotted independently, a valuable capability for studying the complex kinetics of MIEC microelectrodes.

Table I. Numerical parameters for simulation of thin film patterned LSM electrode.

Parameter	Value	Unit	Description
k_{TPB}^{0}	1.0×10^{-11}	mol/(m·s)	TPB reaction rate constant
k_{ads}^{0}	1.0×10^{0}	mol/(m²·s)	Adsorption rate constant
k_{inc}^{0}	6.5×10^{-7}	mol/(m²·s)	Incorporation rate constant
u_{O^-}	1.0×10^{-12}	mol·m²/(J·s)	Adsorbed surface oxygen mobility
D_{O^-}	8.5×10^{-9}	m²/s	Adsorbed surface oxygen diffusivity
θ_0	2.0×10^{-3}	-	Surface site occupancy fraction
Γ	1.0×10^{-6}	mol/m²	Surface site density
$u_{v,m}$	2.6×10^{-14}	mol·m²/(J·s)	Bulk oxygen vacancy mobility
$c_{v,m}^{0}$	1.4×10^{-5}	mol/m³	Equilibrium bulk oxygen vacancy concentration
T	1023	K	Temperature
t	100	nm	Film thickness
w	48	µm	Active film width

To investigate the effect of oxygen partial pressure, we also applied this numerical technique to the case of patterned platinum electrodes for which experimental results have been reported in the literature.[7] The simplification to the case of platinum allows the TPB pathway to be considered without the interference of the bulk pathway, thus helping us to validate the model.

The equilibrium surface coverage, θ_0, was determined by thermodynamic equilibrium considerations for singly-charged, dissociated chemisorbed oxygen (Equation 6),[8] where $\Delta G_{ads,chem}^{0}$ is the standard free energy of reaction and χ_{ms}^{0} is the electrostatic potential difference across the platinum-air surface determined using the parallel-plate capacitor approximation.[4]

$$\frac{1-\theta_0}{\theta_0} P_{O_2}^{1/2} \exp\left(-\frac{F}{RT}\chi_{ms}^{0}\right) = \exp\left(\frac{\Delta G_{ads,chem}^{0}}{RT}\right) \tag{6}$$

Since a change to equilibrium partial pressure should be accompanied by a corresponding change to the equilibrium surface coverage as well as to the exchange current density of pertinent reactions, we adjusted the rate constants in our kinetic model accordingly. As an example, the rate of forward reaction for the TPB process at equilibrium is given in Equation 7[4], where χ_{em} is the potential difference between the platinum and electrolyte

$$\vec{r}_{TPB}^{\,0} = k_{TPB}^{0} = \vec{k}_{TPB}^{\,m} \exp\left(\frac{2F\alpha_{em}}{RT}\chi_{em}\right)\exp\left(\frac{F\alpha_{ms}}{RT}\chi_{ms}^{0}\right)\cdot\theta_0\cdot c_{v,e}^{0}\cdot c_{e',m}^{0} \quad . \tag{7}$$

Let this rate be compared to some reference equilibrium rate, specifically the one occurring with a reference P_{O2} of 0.21 atm, denoted by the superscript or subscript "00." Re-expressing Equation 7 in terms of the reference equilibrium state and assuming no change to χ_{em}, then

$$\vec{r}_{TPB}^{\,0} = k_{TPB}^{0} = \vec{k}_{TPB}^{\,m} \exp\left(\frac{2F\alpha}{RT}\chi_{em}\right)\exp\left(\frac{F\alpha_{ms}}{RT}(\chi_{ms}^{0}-\chi_{ms}^{00})\right)\exp\left(\frac{F\alpha_{ms}}{RT}\chi_{ms}^{00}\right)\cdot\theta_{00}\cdot\left(\frac{\theta_0}{\theta_{00}}\right)\cdot c_{v,e}^{00}\cdot\left(\frac{c_{v,e}^{0}}{c_{v,e}^{00}}\right)\cdot c_{e',m}^{00}\cdot\left(\frac{c_{e',m}^{0}}{c_{e',m}^{00}}\right). \tag{8}$$

Grouping reference parameters together and rearranging,

$$\vec{r}_{TPB}^{\,0} = k_{TPB}^{0} = k_{TPB}^{00}\left[\exp\left(\frac{F\alpha_{ms}}{RT}(\chi_{ms}^{0}-\chi_{ms}^{00})\right)\cdot\left(\frac{\theta_0}{\theta_{00}}\right)\cdot\left(\frac{c_{v,e}^{0}}{c_{v,e}^{00}}\right)\cdot\left(\frac{c_{e',m}^{0}}{c_{e',m}^{00}}\right)\right] \tag{9}$$

where $k_{TPB}^{00} = \vec{k}_{TPB}^{\,m}\exp\left(\frac{2F\alpha}{RT}\chi_{em}\right)\exp\left(\frac{F\alpha_{ms}}{RT}\chi_{ms}^{00}\right)\cdot\theta_{00}\cdot c_{v,e}^{00}\cdot c_{e',m}^{00}$ is the reference equilibrium rate

constant. Thus, one equilibrium rate constant may be specified for a reference case and adjusted to different equilibrium partial pressures using the term in brackets in Equation 9 along with additional equilibrium data, specifically θ_0 provided by Equation 6. Both the TPB and adsorption rate constants were adjusted in this manner.

We simulated transport and reaction rate on the 1D platinum surface and 2D solid oxide electrolyte using a finite volume discretization solved with Newton's method programmed in Matlab. Both steady state and alternating current (AC, to be described later in this paper) calculations were performed. The simulated AC response at different values of oxygen partial pressure is given in Figure 3 along with the experimentally reported values.[7] In the figure, the charge transfer resistivity, ρ_{ct}, is determined from the width of the resulting impedance loop normalized by the total length of TPB line of the microelectrode array. Numerical parameters used for the simulation are listed in Table II. The simulated results agree quite well with the experimental data.

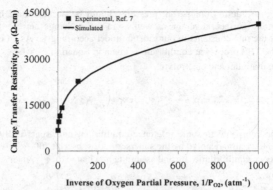

Figure 3. Charge transfer resistivity, ρ_{ct}, of an array of platinum microelectrodes with TPB density of 250 cm/cm^2 and temperature of 750°C at various oxygen partial pressures, P$_{O2}$. Experimental results from reference 7 are shown with our simulation results.

Table II. Numerical parameters for platinum-yttria stabilized zirconia (YSZ) TPB simulation. Reference state refers to P$_{O2}$=0.21 atm.

Parameter	Value	Unit	Description
$k_{TPB}^{\,00}$	2.3×10^{-9}	mol/(m·s)	Reference TPB reaction rate constant
$k_{ads}^{\,00}$	1.0×10^{2}	mol/(m^2·s)	Reference adsorption reaction rate constant
$\Delta G_{ads,chem}^{\,0}$	-5.4×10^{4}	J/mol	Standard free energy of adsorption reaction
u_{O^-}	1.0×10^{-12}	mol·m^2/(J·s)	Adsorbed surface oxygen mobility
D_{O^-}	8.5×10^{-9}	m^2/s	Adsorbed surface oxygen diffusivity
Γ	1.0×10^{-6}	mol/m^2	Surface site density
σ_{YSZ}	2.1×10^{0}	S/m	Conductivity of YSZ
L_{TPB}	250	cm/cm^2	Density of TPB line
T	1023	K	Temperature

LOCAL ELECTRICAL AND CHEMICAL STATES

In addition to simulating reaction rates on the local level and summing for overall cell response, the numerical thin film model allows insight into the local electrical and chemical states of the MIEC bulk and surface as well as the solid oxide electrolyte bulk. This type of information is not experimentally trivial to obtain and modeling provides a convenient way to access it. Knowing the local electrical and chemical states within the MIEC and electrolyte aids the understanding of thin film patterned electrode experiments

Electrical states

Sheet resistance occurs inside porous as well as patterned MIEC electrodes, adding another degree of difficulty into the interpretation of experimental results and an additional impediment to experimental and practical design. Our numerical continuum model was designed to incorporate the effect of sheet resistance[3] in order to help understand and control this often-neglected factor. Figure 4a shows a simulated potential profile across the width of a patterned electrode (see geometry in Figure 1).

As the distance from the current collector increases, the potential increasingly varies from the applied value. This deviation can impact both the bulk as well as the TPB kinetics substantially.

Figure 4b shows the simulated potential profile due to Ohmic drop within the electrolyte. The local potential at the TPB is important to determining the overall electrochemical response and is coupled to the other factors. Therefore, it must be determined numerically and simultaneously with the rest of the model.

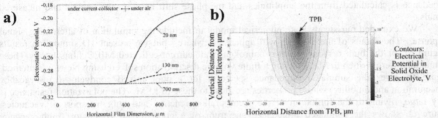

Figure 4. a) Simulated electrical potential across the width of a patterned MIEC electrode due to sheet resistance for different values of film thickness. The effect becomes more severe as the aspect ratio of the film becomes more extreme, which can affect the bulk and TPB kinetics. Cell voltage was 0.300V (cathodic bias). See Table I for numerical parameters except k_{TPB}^0, which was 1×10^{-12} mol/(m·s). b) Simulated electrical potential in the electrolyte due to the flow of oxygen vacancies to the TPB. The local electrical potential of the solid electrolyte is one factor that can significantly affect the TPB kinetics. Cell voltage was 0.100 V (cathodic bias). See Table I for numerical parameters, except $k_{TPB}^0=1\times10^{-12}$, $k_{ads}^0=3.0\times10^{-5}$.

Chemical States

Similarly, the concentration of chemical species such as point defects, critical to the overall response, can be simulated and imaged (Figure 5). Such a picture allows understanding of how the reactants and products are distributed within the solid-state system. With this understanding, better design of electrodes can be achieved. Varying degrees of complication in terms of defect chemistry can be employed to balance the accuracy of results versus computation time.

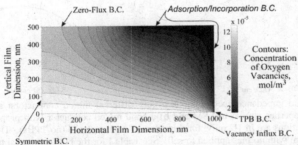

Figure 5. Simulated distribution of oxygen vacancies within a patterned MIEC. The concentration distribution depends upon the rate of vacancy influx from the electrolyte, the rate of consumption on

air-exposed surfaces, and the local electrical field. The location of the various boundary conditions (B.C.) are labeled. In particular, vacancies are consumed by the incorporation B.C.

AC SIGNAL RESPONSE

Electrochemical impedance spectroscopy is one of the most important experimental tools in electrochemistry. It involves the application of an alternating voltage (or current) to an electrochemical system and the measurement of the resulting alternating current (or voltage). Impedance is calculated from the amplitudes and the phase shift between the applied and measured signals.

We extended our steady state numerical model to include the simulation of alternating signal response. The details of the analysis will appear elsewhere,[9] but we present 1D simulation results (Figure 6) for the bulk response of two unbroken, well current-collected MIEC films here. These simulations were written in Matlab using a finite volume discretization solved using Newton's method. Figure 6a gives the simulated impedance spectrum for a poor mixed conductor (low vacancy concentration and mobility) whose properties are close to those of LSM. The polarization resistance is very large, owing to the limitation of the incorporation reaction and bulk transport of vacancies. Figure 6b shows the simulated impedance spectrum of a good mixed conductor (high vacancy concentration and mobility) thin film, whose properties are close to those of the family $La_{1-x}Sr_xCo_{1-y}Fe_yO_{3-\delta}$ (LSCF). Two processes occurring on very different time scales are apparent. The high-frequency process corresponds to vacancy transfer over the MIEC-electrolyte interface in parallel with capacitive charging of the same interface. The low-frequency feature corresponds to processes occurring on/in the MIEC, with resistance related to surface kinetics and capacitance related to changes in the through-thickness bulk defect concentration. Our simulation methodology allows for the seamless transition between cases a) and b), two extremes in electrochemical response, simply by changing the parameters of the continuum simulation; there are no required changes in assumptions or use of equivalent circuits. It should be noted that the relations between kinetic processes and the resulting impedance spectra are nontrivial and constitute an additional tool for investigation and prediction of impedance spectra. The numerical parameters for the poor MIEC film are listed in Table I, while the parameters for the good film are listed in Table III.

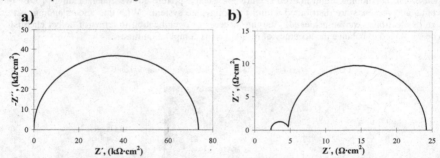

Figure 6. Simulated impedance spectra for two very different mixed conductors: a) a poor mixed conductor with properties similar to LSM and b) a good mixed conductor with properties not unlike LSCF. The simulations are in 1D only, for thin films of each material (no TPBs). The film thickness in each case is 1 μm with no bias applied to cell. See Table I for the bulk numerical parameters of the poor mixed conductor. The bulk numerical parameters of the good mixed conductor are listed in Table III. In both cases, the capacitance along the MIEC-electrolyte interface is 5 F/m².

Table IIII. Numerical parameters for AC simulation of good mixed conductor thin film.

Parameter	Value	Unit	Description
$k_{ads}^{\ 0}$	1.0×10^0	mol/(m²·s)	Adsorption rate constant
$k_{inc}^{\ 0}$	1.0×10^{-4}	mol/(m²·s)	Incorporation rate constant
θ_0	2.0×10^{-3}	-	Surface site occupancy fraction
Γ	1.0×10^{-6}	mol/m²	Surface site density
$u_{v,m}$	1.0×10^{-13}	mol·m²/(J·s)	Bulk oxygen vacancy mobility
$c_{v,m}^{\ 0}$	1.0×10^3	mol/m³	Equilibrium bulk oxygen vacancy concentration
T	1023	K	Temperature
C_{em}	5.0×10^0	F/m²	Capacitance along MIEC-electrolyte interface

EXPERIMENTAL DESIGN

One of the key aspects of our model is that it can be used to inform experiments on real materials. For example, it can be used to explore the effect of sheet resistance and also to predict which pathways will be important and worth maximizing in design of porous electrodes. This knowledge can be used to guide actual experiments on electrodes of well-controlled geometry to verify the predictions. By extending our model to the cases described in this contribution, we have added additional capabilities and expanded the range of experiments to which our modeling is relevant.

CONCLUSION

We outlined our progress in extending our existing numerical modeling technique for thin film MIECs in several experimentally relevant ways. We added the consideration of the TPB reaction and accompanying surface transport. We highlighted the additional information, namely local electrical and concentration fields, available from the extension. Finally, we extended our approach to the case of alternating signal simulation for eventual comparison to electrochemical impedance experiments. The extension of our model allows us to address important aspects in the investigation of real MIEC materials in well-controlled geometries.

ACKNOWLEDGEMENT

The authors thank Prof. Yingjie Liu and Prof. Tom Fuller for helpful advice and conversations. This work was supported by DOE-SECA Core Technology Program under grant No. DE-NT-0006557.

REFERENCES
1. D.S. Mebane and M. Liu, Classical, Phenomenological Analysis of the Kinetics of Reactions at the Gas-Exposed Surface of Mixed Ionic Electronic Conductors, *J. Solid State Electrochem.*, **10**, 575-80 (2006).
2. D.S. Mebane and M. Liu, Classical, Phenomenological Analysis of the Kinetics of Reaction at the Gas-Exposed Surface of Mixed Ionic Electronic Conductors: Erratum, *J. Solid State Electrochem.*, **11**, 448 (2007).
3. D.S. Mebane, Y. Liu, and M. Liu, A Two-Dimensional Model and Numerical Treatment for Mixed Conducting Thin Films: The Effect of Sheet Resistance, *J. Electrochem. Soc*, **154**, A421-A26 (2007).
4. M.E. Lynch, D.S. Mebane, Y. Liu, and M. Liu, Triple Phase Boundary and Surface Transport in Mixed Conducting Patterned Electrodes, *J. Electrochem. Soc*, **155**, B635-B43 (2008).
5. E. Siebert, A. Hammouche, and M. Kleitz, Impedance Spectroscopy Analysis of La$_{1-x}$Sr$_x$MnO$_3$-Yttria-Stabilized Zirconia Electrode-Kinetics, *Electrochim. Acta*, **40**, 1741-53 (1995).

6. H. Lauret and A. Hammou, Localization of Oxygen Cathodic Reduction Zone at Lanthanum Manganite Zirconia Interface, *J. Eur. Ceram. Soc.*, **16**, 447-51 (1996).
7. R. Radhakrishnan, A.V. Virkar, and S.C. Singhal, Estimation of Charge-Transfer Resistivity of Pt Cathode on Ysz Electrolyte Using Patterned Electrodes, *J. Electrochem. Soc*, **152**, A927-A36 (2005).
8. J. Fleig, R. Merkle, and J. Maier, The P(O-2) Dependence of Oxygen Surface Coverage and Exchange Current Density of Mixed Conducting Oxide Electrodes: Model Considerations, *Phys. Chem. Chem. Phys.*, **9**, 2713-23 (2007).
9. M.E. Lynch, D.S. Mebane, and M. Liu, in preparation,

LAMINAR FLOW AND TOTAL PRESSURE EFFECTS IN SOLID OXIDE FUEL CELL ELECTRODE PORES AND THEIR EFFECTS ON VOLTAGE-CURRENT CHARACTERISTICS

V. Hugo Schmidt, R. R. Chien, and Laura M. Lediaev
Department of Physics, Montana State University
Bozeman, MT 59717, USA

ABSTRACT

This work describes binary and ternary gas flow in solid oxide fuel cell electrode pores more accurately by considering laminar flow and avoiding the approximation that total pressure is uniform along the pores. The resulting voltage-current characteristics are found for a particular cell design for both the fuel cell and electrolysis modes. The equations for these characteristics clearly separate the terms that affect activation polarization and concentration polarization. In both terms, the light mass of the H_2 molecule increases the power output relative to other fuels.

Our analysis begins with the diffusion equation for each gas species in tortuous pores with assumed circular cross-section. Each equation includes a Knudsen term for wall collisions, an unlike-molecule collision term, and a laminar flow term.

Without including the laminar flow term, for binary flow, such as H_2 and H_2O in anode pores, total pressure increases linearly across the anode, but fuel and exhaust partial pressures have linear and quadratic terms. For typical pore diameters near one micron, the Knudsen and intermolecular collisions have comparable importance. For ternary flow (fuel and diluent input plus exhaust gas), each partial pressure has an added exponential term. Including laminar flow gives terms of all orders for the total and partial pressures.

We find that for practical electrode designs, we must keep terms to first order in total as well as partial pressure dependence on position across the electrode, but higher order terms are negligible and effects of laminar flow are small.

INTRODUCTION

Solid Oxide Fuel Cells (SOFCs) provide an efficient source of electricity, and are free of pollutants and CO_2 if the fuel is hydrogen. They can also be run in the reverse Solid Oxide Electrolysis Cell (SOEC) mode to produce hydrogen from steam, for instance. To analyze their operation, one must consider a number of cell construction parameters and operating conditions. Cell construction parameters include anode and cathode materials, porosity, tortuosity, and thickness, as well as electrolyte material and thickness, and conductivity as a function of temperature. The electrolyte conductivity is typically from oxygen ion vacancies as is assumed in this work, but can be protonic, may have an unwanted electronic component, or may be some combination of these. A crucial cell construction parameter is the electrolyte-electrode interface, including the length of Triple Phase Boundary (TPB) per unit electrolyte cross-sectional area.

The operating conditions include temperature, total pressure at the electrode outer surfaces, input flow rates of fuel and oxidant gases and any other gases such as N_2 in air input, and an electrical condition. Current density i in the electrolyte is more convenient for mathematical analysis than terminal voltage V as an electrical condition, and output power W is unsatisfactory because two (V, i) values correspond to each W value except for W_{peak}.

Given a current density i as a specified operating condition, the flow rates of the gases in the electrodes are known immediately. Then the gas partial pressures at the electrode outer surfaces can be calculated because the input gas flow rates and total pressure are known. One main purpose of this paper is to calculate gas partial pressures as functions of position in the electrodes from these outer surface partial pressures, the gas flow rates in the electrodes, and the electrode parameters.

139

Once these position-dependent partial pressures are known, the partial pressures at the electrolyte-electrode interfaces are known. Using our model for the reactions at these interfaces, the voltage steps across these interfaces can be calculated, and then after subtracting the ohmic drop across the electrolyte, the SOFC or SOEC terminal voltage can be calculated. Such reaction models and terminal voltage calculations constitute a second main purpose of this paper.

Spatial dependences of partial pressures are determined by three physical processes, namely intermolecular collisions of unlike molecules, molecular collisions with electrode pore walls (Knudsen diffusion), and laminar flow. Laminar flow is emphasized in this paper because it is neglected in some previous works, as is the variation of total pressure across the electrodes. Indeed, laminar flow is absent only if pressure is uniform across the electrode. Even for reactions for which there are equal and opposite flow rates of numbers of molecules per second, there will in general be both laminar flow and total pressure variation with position.

MODELING AND CALCULATIONS

We have developed a first-principles model for the $V(i)$ (terminal voltage vs. current density) curve for SOFC's that extends across the SOEC (Solid Oxide Electrolyzer Cell) range as well, where an $i(V_{act})$ expression that predicts the correct limiting current density was derived.[1-3] The Butler-Volmer equation for current density i as a function of activation polarization V_{act} predicts that i increases without limit for large applied voltage in the SOEC mode. Our model predictions were in accord with experiments in other labs for operation both in the SOFC and SOEC modes. In addition, that model was extended to include predictions of performance at elevated pressures.[4]

Here we extend the model to include laminar flow. Laminar flow is considered for gas flow through the anode to find the new concentration polarization for the binary and ternary cases, meaning the presence of two or three gases in the anode. The results are compared with those obtained from our previous calculations[2,4] that did not consider laminar flow. We start with the Stefan-Maxwell equation including molecular and Knudsen diffusion in the pores, including a laminar flow term from the Darcy permeation model. In molar units it is

$$\frac{N_i}{D_{K,i}} + \sum_{j=1, j\neq i}^{n} \frac{X_j N_i - X_i N_j}{D_{ij}} = \frac{-1}{RT}\left[\frac{d(PX_i)}{dx} + \frac{B_o P}{\eta D_{K,i}} X_i\left(\frac{dP}{dx}\right)\right], \tag{1}$$

where N_i and N_j are molar fluxes of components i and j (mol/cm²s), respectively, D_{Ki} and D_{ij} are the Knudsen diffusion coefficient for component i and the binary diffusion coefficient for components i and j, respectively, X_i and X_j are the fractional molar concentrations of components i and j, P is the total pressure, R is the gas constant, and T is the absolute temperature. We emphasize that x is the coordinate along a typical anode diffusion path, not straight across the anode. B_o is the flow permeability, which can be determined by the Kozeny–Carman relationship[5,6] as

$$B_o = \frac{(2\bar{r})^2}{72}\left(\frac{w}{L_e}\right)^2 \frac{\phi^3}{(1-\phi)^2}, \tag{2}$$

where it is assumed that the porous electrode is formed by closely packed spherical particles with diameter $(=2\bar{r})$, which is certainly an idealization. w is the width of the anode and L_e is the actual length of a typical tortuous flow path along the anode pores. $\tau = L_e/w$ is tortuosity and ϕ is porosity. It also can be expressed as

$$B_o = \frac{\phi^3}{72\tau^2(1-\phi)^2}(2\bar{r})^2. \tag{3}$$

Viscosity η of the gas mixture can be determined by the method of Herning and Zipperer[7],

$$\eta_{mix} = \frac{\sum_{i=1}^{n}\left(X_i \eta_i M_i^{1/2}\right)}{\sum_{i=1}^{n}\left(X_i M_i^{1/2}\right)}, \tag{4}$$

where X_i, η_i, and M_i are mole fraction, viscosity of component i, and molecular weight of component i. The dynamic viscosity of component i, (η_i) can be determined by the sixth-order polynomial functions developed by Todd and Young[8] after averaging viscosities of gases based on their mole fractions,

$$\eta(\mu P) = \sum_{k=0}^{6} b_k \gamma^k, \tag{5}$$

where $\gamma = T(K)/1000$, μP is the unit of viscosity, i.e. the micro-poise (1 $\mu P = 10^{-7}$ kg/m-s), and the b_k are numerical coefficients.

INTEGRATION OF GAS FLOW INTO $V(i)$ EXPRESSION

In previous work,[2,3] we included the effect of limitation on gas flow (the concentration polarization effect) on the $V(i)$ characteristic of an SOFC or SOEC only in an approximate manner. In Ref. 2, we found the tortuosity τ for the Jiang and Virkar[9] SOFC anode in their anode-supported cell for which they and we neglect flow limitations. In Ref. 3, we developed a model for the anode and cathode reactions and the resulting $V(i)$ expression, without explicitly including some anode parameters in the model. Now we include all these parameters, including the known τ, for the binary flow case without considering laminar flow. The resulting expression clearly separates flow effects (concentration polarization) from reaction effects (activation polarization). We will show how the numerical $V(i)$ results change due to inclusion of laminar flow effects. Related work on gas transport in SOFC anodes has appeared.[10-13]

We start with the $V(i)$ expression we developed previously and compared with $V(i)$ expressions by others.[3] Unlike these others, ours has a symmetric form

$$V = U/q - (kT/q)\ln[(b+i)(d+i)^{1/2}/(a-i)(c-i)^{1/2}] - V_{ohm}. \tag{6}$$

This equation holds for reaction pairs such as $H_2 + O^{2-} \leftrightarrow H_2O + 2e^-$ at the anode and $O_2 + 4e^- \leftrightarrow 2O^{2-}$ at the cathode, for which equal numbers of fuel and exhaust molecules take part in the anode reaction, and for which twice the charge is transferred in the cathode reaction. It holds also for the CO/CO_2 fuel/exhaust pair for the anode reaction. In Eq. (6), U is the energy of reaction, q is the $2e$ charge transfer for the anode reaction, k is Boltzmann's constant, T is temperature, i is current density in the electrolyte, and a, b, c, d are attempt current densities. The $\frac{1}{2}$ powers occur because the cathode reaction has charge transfer $2q$. The attempt current densities a and b at the anode-electrolyte interface, multiplied by their respective reaction success probabilities, are the actual current densities, which for open circuit become equal and are called the anode exchange current density. Similarly, c and d are the cathode attempt current densities. The cathode exchange current density in general will differ from its anode counterpart.

The form of Eq. (6) is related to the reaction success probabilities, as explained previously.[3] Here we only discuss briefly the forms of a, b, c, d, because their factors occur in the detailed expression for V that we will derive. For the anode forward (fuel $\rightarrow$ exhaust) attempt current density a we have

$$a = \frac{1}{2} n_{fw} (kT/m_f)^{1/2} q(1-\upsilon)f, \tag{7}$$

where n_{fw} is the fuel gas molecular concentration at the anode-electrolyte interface, a distance w across the anode from the anode outer surface which is in contact with the anode gas plenum. The fuel molecule mass is m_f, and $(kT/m_f)^{1/2}$ is the average impingement speed of the $\frac{1}{2}$ of the fuel molecules that are moving toward the interface. The forward reaction is assumed to be possible only if the fuel molecule impinges on a non-vacant oxygen ion site on the electrolyte surface, and also that site must lie on a Triple Phase Boundary (TPB) where anode material, an anode pore, and electrolyte meet. The fraction of oxygen ions sites that are vacant is designated as υ, and f represents the fraction of oxygen ion sites adjacent to a TPB as estimated by a method presented previously.[3]

For the anode reverse (exhaust$\rightarrow$fuel) attempt current density b, Eq. (7) is changed only by changing the subscript $_f$ to $_e$ and changing $(1-\upsilon)$ to υ, so we obtain

$$b=\tfrac{1}{2}n_{ew}(kT/m_e)^{1/2}q\upsilon f. \tag{8}$$

For the cathode forward (oxygen ionization) attempt current density c, Eq. (8) is changed by changing q to $2q$, and by squaring υ because two adjacent vacant oxygen ion sites are needed for the oxygen molecule to adhere strongly to the electrolyte surface. Also, we must change the subscript $_e$ to $_c$, and omit the subscript $_w$ because in this analysis we are considering anode-supported SOFCs and are neglecting flow impedance of the cathode. Accordingly, c has no dependence on i and can be designated as its open-circuit value c_0, because none of its factors have i dependence, so c has the form

$$c=c_0=\tfrac{1}{2}n_c(kT/m_c)^{1/2}(2q)\upsilon^2 f. \tag{9}$$

Specifically, n_c just corresponds to the 21% oxygen concentration in air at atmospheric pressure. We assume for simplicity that υ has no i dependence and is the same at the anode and cathode interfaces, because we so far have no model for such dependence, which could in principle exist.

For the cathode reverse (oxygen molecule creation from oxygen ions) attempt current density d, we have an entirely different expression as discussed previously.[3] This expression is

$$d=d_0=8q\nu(1-\upsilon)^2 f/s, \tag{10}$$

where ν is the attempt frequency for a pair of adjacent oxygen ions on the surface to eject four electrons and form an oxygen molecule, and $s^{1/2}=2.62\times10^{-8}$ cm is the separation distance of adjacent oxygen ions which form a square array on a (001) surface of the yttria-stabilized zirconia (YSZ) electrolyte surface. As assumed for c, we assume d has no i dependence, so it can be designated as the open-circuit value d_0. Furthermore, $d_0>>i$ by about 5 orders of magnitude for any achievable i values, so $(d+i)^{1/2}$ in Eq. (6) can be replaced by $d_0^{1/2}$ as an excellent approximation.

The next step is factoring the open-circuit values of a, b, c, d out of the factors in parentheses in the ln argument in Eq. (6). Then, for $i=0$, we have the open-circuit voltage

$$V_0=U/q-(kT/q)\ln(b_0d_0^{1/2}/a_0c_0^{1/2}). \tag{11}$$

This V_0 expression is considerably smaller than U/q because d_0 is about 5 orders of magnitude larger than a_0, b_0, and c_0. We found previously[3] that $U/q=1.214$ V.

For nonzero i we then have

$$V=V_0-(kT/q)\ln[(b/b_0+i/b_0)/(a/a_0-i/a_0)(1-i/c_0)^{1/2}]-V_{ohm}. \tag{12}$$

The final task is finding a_0, b_0, and the i dependences of a and b. This is done by using expressions in our previous work[2] which was aimed at finding the tortuosity of the Jiang and Virkar[9] SOFC. In Eqs. (7) and (8) respectively, n_{fw} and n_{ew} are power series in i ending at the i^2 terms. The corresponding a/a_0 and b/b_0 expressions are similar power series in i, for which the constant terms are unity. Eq. (12) then becomes

$$V=V_0-(kT/q)\ln[(1+i/B_1+i^2/B_2^2)/(1-i/A_1-i^2/A_2^2)(1-i/c_0)^{1/2}]-V_{ohm}. \tag{13}$$

In this expression, c_0 is given in Eq. (9), and the other coefficients are

$$A_1 = p_f i_1/[1+C_{aAP}+C_{aCP}] = p_f i_1/[1+2i_1/n_1(kT/m_f)^{1/2}q(1-\upsilon)f+(1/D_{Kf}+1/D_{efl})\tau^2 w i_1/n_1\phi q], \tag{14}$$

$$B_1 = (1-p_f)i_1/[1+C_{bAP}+C_{bCP}] = (1-p_f)i_1/[1+2i_1/n_1(kT/m_e)^{1/2}q\upsilon f+(1/D_{Ke}+1/D_{efl})\tau^2 w i_1/n_1\phi q], \tag{15}$$

$$A_2 = B_2 = (n_1 q\phi/\tau^2 w)\{2(1-p_f)D_{Kf}D_{efl}/[(m_e/m_f)^{1/2}-1]\}^{1/2}. \tag{16}$$

The C's are dimensionless constants whose subscripts $_{AP}$ and $_{CP}$ refer respectively to their giving rise to. to Activation Polarization and Concentration Polarization effects on V, and the $_a$ and $_b$ subscripts refer to the forward and reverse anode reaction factors respectively in the ln argument in Eq. (13).

We now define parameters in Eqs. (14-16) that have not been defined previously. The molecular fraction of fuel gas in the metered input gases is p_f. For the Jiang and Virkar[9] SOFC runs, in the parameter $i_1 = j_1 q/S$, j_1 is the total metered input gas flow rate of 140 ml/min which corresponds to 6.286×10^{19} molec/s, and S is the electrolyte cross-sectional area of 1.1 cm^2. Accordingly, $i_1 = 18.49$ A/cm^2 would be the current density if all metered input gas were fuel, and if 100% fuel utilization were achieved. Jiang and Virkar used $P_1 = 1$ atm pressure at the anode and cathode outer surfaces and ran their cell at 1073 K, so the total gas concentration at these surfaces was $n_1 = P_1/kT = 6.84 \times 10^{18}$/cm^3.

The remaining undefined parameters appear in C_{aCP} and C_{bCP} and are the Knudsen or wall collision diffusion constants D_{Kf} and D_{Ke} for the fuel and exhaust gas respectively, the intermolecular collision diffusion constant D_{efl} where the subscript $_l$ indicates that this is its value for 1 atm total pressure, the tortuosity τ, the anode thickness w, and the anode porosity ϕ. Note that the Knudsen constants are proportional to pore radius of 0.5 micrometer, and inversely proportional to the square root of molecular mass.

For the binary case, with a 50/50% molecular ratio of H_2/H_2O for the metered input gas, chosen because it can best illustrate operation in both the SOFC and SOEC modes, and 21% O_2 in the air on the cathode side, and operating temperature 1073 K and pressure 1 atm on the electrode outer surfaces, the numerical values of the above constants in Eqs. (9) and (14)-(16) are

$$p_f = 0.5,\ i_1 = 18.29\ \text{A/cm}^2,\ D_{Kf} = 11.3\ \text{cm}^2/\text{s},\ D_{Ke} = 3.767\ \text{cm}^2/\text{s},\ D_{efl} = 7.704\ \text{cm}^2/\text{s},\ \tau = 2.3, \tag{17}$$

$w = 0.11$ cm, $n_1 = 6.855 \times 10^{18}$/cm^3, $\phi = 0.54$, $q = 3.2 \times 10^{-19}$ C, $k = 1.38 \times 10^{-23}$ J/K, $T = 1073$ K, $m_f = 3.32 \times 10^{-27}$ kg, $m_e = 29.88 \times 10^{-27}$ kg, $m_c = 53.12 \times 10^{-27}$ kg, $n_c = 1.440 \times 10^{18}$/cm^3, $\upsilon = 0.7$, $f = 4.22 \times 10^{-4}$, $v = 1.0 \times 10^{13}$/s, $a_0 = (29.32\ \text{A/cm}^2)p_f = 14.66$ A/cm^2, $b_0 = (22.81\ \text{A/cm}^2)(1-p_f) = 11.40$ A/cm^2, $c_0 = 5.03$ A/cm^2, $d_0 = 1.42 \times 10^6$ A/cm^2, $V_0 = 0.935$ V.

Of the above, only υ is an arbitrary constant, because presently we have no way to determine its value, or whether it depends on i. The V_{ohm} term in the above equations for V is given by $V_{ohm} = (0.122$ ohm-cm$^2)i$, based on results presented by Jiang and Virkar.[9]

NEWTON'S METHOD / FINITE ELEMENT METHOD SOLUTION FOR GAS FLOW INCLUDING LAMINAR FLOW

To find n_{fw} and n_{ew} in Eqs. (7) and (8), we solve equations derived from Eq. (1) but expressed in molecular units. These are coupled nonlinear equations which describe the concentration (molecules per volume) as a function of distance along the anode. In the binary case, where n_1 (in this Section only) designates the fuel (H_2) concentration, and n_2 is exhaust (H_2O) concentration, we have Eqs. (18) and (19) below. The last term in each equation is the laminar flow term, which includes viscosity (η) effects, and x is distance along a typical pore path through the anode.

$$\frac{J_1}{D_1} - \frac{J_2}{D_{12}}n_1 + \frac{J_1}{D_{12}}n_2 + \frac{\partial n_1}{\partial x} + \frac{\alpha}{D_1}\frac{n_1}{\eta}\left(\frac{\partial n_1}{\partial x} + \frac{\partial n_2}{\partial x}\right) = 0 \tag{18}$$

$$\frac{J_2}{D_2} + \frac{J_2}{D_{12}}n_1 - \frac{J_1}{D_{12}}n_2 + \frac{\partial n_2}{\partial x} + \frac{\alpha}{D_2}\frac{n_2}{\eta}\left(\frac{\partial n_1}{\partial x} + \frac{\partial n_2}{\partial x}\right) = 0 \tag{19}$$

$$\alpha \equiv B_0 k_B T, \quad D_{ij} \equiv \frac{P_1}{k_B T}D_{ij1}, \quad \frac{1}{\eta} = \frac{n_1\sqrt{m_1} + n_2\sqrt{m_2}}{n_1\eta_1\sqrt{m_1} + n_2\eta_2\sqrt{m_2}} \tag{20}$$

In the special case where laminar flow is neglected, and $J_2=-J_1$, we get a quadratic solution for the concentrations. Also, in general for any number of molecular species we always have that the total concentration $(n_1 + n_2 + ...)$ is a linear function if laminar flow is neglected.

$$n_1(x) = n_{10} - J_1\left(\frac{1}{D_1} + \frac{n_{10} + n_{20}}{D_{12}}\right)x + \frac{J_1^2}{2D_{12}}\left(\frac{1}{D_1} - \frac{1}{D_2}\right)x^2 \tag{21}$$

$$n_2(x) = n_{20} + J_1\left(\frac{1}{D_2} + \frac{n_{10} + n_{20}}{D_{12}}\right)x - \frac{J_1^2}{2D_{12}}\left(\frac{1}{D_1} - \frac{1}{D_2}\right)x^2 \tag{22}$$

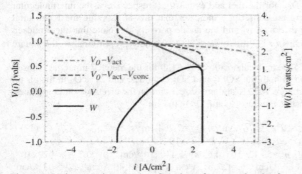

Figure 1. Voltage and power as a function of current density for anode thickness $= 0.11$ cm.

In order to solve the laminar flow case, which is nonlinear, we use Newton's method as the iterative solver, with the linear solution as the initial guess. To discretize the problem, we used the Finite Element Method. In order to use the FEM, the differential equations are put into a polynomial form, so that the unknown variables do not appear in the denominator.

$$\left[\frac{J_1}{D_1} - \frac{J_2}{D_{12}}n_1 + \frac{J_1}{D_{12}}n_2 + n_1'\right](n_1\eta_1\sqrt{m_1} + n_2\eta_2\sqrt{m_2}) + \frac{\alpha}{D_1}n_1\left(n_1' + n_2'\right)(n_1\sqrt{m_1} + n_2\sqrt{m_2}) = 0 \tag{23}$$

$$\left[\frac{J_2}{D_2} + \frac{J_2}{D_{12}}n_1 - \frac{J_1}{D_{12}}n_2 + n_2'\right](n_1\eta_1\sqrt{m_1} + n_2\eta_2\sqrt{m_2}) + \frac{\alpha}{D_2}n_2\left(n_1' + n_2'\right)(n_1\sqrt{m_1} + n_2\sqrt{m_2}) = 0 \tag{24}$$

In the Finite Element Method the unknown variables are represented by a sum of coefficients times basis functions. Each basis function is nonzero only inside its own element.

$u(x) \approx \sum_j u_j \psi_j(x)$, where u = n_1 or n_2 (25)

The substitution is made in the differential equations, and then those equations are multiplied by the ith basis function, and integrated over the entire domain. As an example, the converted form of the first differential equation (for n_1) is shown below. We now have a vector, which we want to be uniformly zero. To try to find the solution which gets us to zero, we use Newton's method. At each iteration we need to invert the Jacobian, where $J_{ij} = \partial F_i / \partial u_j$. It only takes a couple of iterations to get to the final solution. As an implementation note, all the integrals need only be done for a single element, and only once. At each iteration the integrals then become a sum over all the elements. For the calculation, we use 15 elements, and cubic Lagrange basis functions.

$$F_i^1 = \int \psi_i \left\{ \left(\eta_1 \sqrt{m_1} \sum_j n_{1j}\psi_j + \eta_2 \sqrt{m_2} \sum_j n_{2j}\psi_j \right) \left(\frac{J_1}{D_1} - \frac{J_2}{D_{12}} \sum_k n_{1k}\psi_k + \frac{J_1}{D_{12}} \sum_k n_{2k}\psi_k + \sum_k n_{1k}\psi'_k \right) \right. $$
$$\left. + \frac{\alpha}{D_1} \sum_j n_{1j}\psi_j \left(\sqrt{m_1} \sum_k n_{1k}\psi_k + \sqrt{m_2} \sum_k n_{2k}\psi_k \right) \left(\sum_l n_{1l}\psi'_l + \sum_l n_{2l}\psi'_l \right) \right\} dx$$
(156)

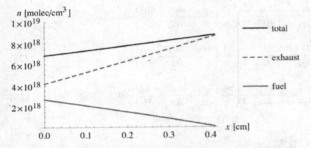

Figure 2. Concentration as a function of pore distance in the anode.

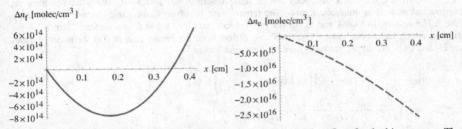

Figure 3. The difference between the solutions with and without laminar flow for the binary case. The difference is less than 1%.

The full solution is remarkably similar to the linear case. Laminar flow has a very small effect on the concentrations (and voltage). The Figures above show the difference between the full solutions and the linear solutions for current density $i=2.0$ A/cm^2. The percent difference is less than one percent, so for practical purposes viscosity can be neglected.

We now discuss some features of the results shown in these Figures. In Fig. 1, the predicted terminal voltage V and power density W are plotted for a 50/50 molecular H_2/H_2O metered input gas ratio for 1 atm and 1073 K operating pressure and temperature, for the cell design reported by Jiang and Virkar.[9] Sharp cutoffs are seen at 2.55 A/cm^2 and -1.71 A/cm^2, due respectively to H_2 starvation (SOFC mode) and H_2O starvation (SOEC mode). The V_0-V_a-V_c curve shows the performance expected if ohmic polarization (electrolyte resistivity) V_{ohm} could be eliminated. The V_0-V_a curve shows the performance expected if also the concentration polarization V_c in the anode of this anode-supported cell could be eliminated. Note that the cutoff at i=5.03 A/cm^2 due to the activation polarization V_a is due to oxygen starvation, and not hydrogen starvation as for the SOFC-mode cutoffs for the other curves.

Fig. 2 shows gas concentrations across the anode. As discussed above, the total concentration is linear and has significant variation across the anode, even for our case of equal and opposite fuel and exhaust molecular fluxes through the anode. The fuel and exhaust concentrations have barely discernible quadratic components in addition to strong linear dependences.

Fig. 3 shows the fuel and exhaust gas concentration changes across the anode due to laminar flow effects. For comparison, the total gas concentration at 1 atm is 6.855x10^{18} molec/cm^3.

Ternary Case

$$\frac{J_1}{D_1} - \left(\frac{J_2}{D_{12}} + \frac{J_3}{D_{13}}\right) n_1 + \frac{J_1}{D_{12}} n_2 + \frac{J_1}{D_{13}} n_3 + \frac{\partial n_1}{\partial x} + \frac{\alpha}{D_1} \frac{n_1}{\eta}\left(\frac{\partial n_1}{\partial x} + \frac{\partial n_2}{\partial x} + \frac{\partial n_3}{\partial x}\right) = 0 \tag{27}$$

$$\frac{J_2}{D_2} + \frac{J_2}{D_{12}} n_1 - \left(\frac{J_1}{D_{12}} + \frac{J_3}{D_{23}}\right) n_2 + \frac{J_2}{D_{23}} n_3 + \frac{\partial n_2}{\partial x} + \frac{\alpha}{D_2} \frac{n_2}{\eta}\left(\frac{\partial n_1}{\partial x} + \frac{\partial n_2}{\partial x} + \frac{\partial n_3}{\partial x}\right) = 0 \tag{28}$$

$$\frac{J_3}{D_3} + \frac{J_3}{D_{13}} n_1 + \frac{J_3}{D_{23}} n_2 - \left(\frac{J_1}{D_{13}} + \frac{J_2}{D_{23}}\right) n_3 + \frac{\partial n_3}{\partial x} + \frac{\alpha}{D_3} \frac{n_2}{\eta}\left(\frac{\partial n_1}{\partial x} + \frac{\partial n_2}{\partial x} + \frac{\partial n_3}{\partial x}\right) = 0 \tag{29}$$

$$\frac{1}{\eta} = \frac{n_1\sqrt{m_1} + n_2\sqrt{m_2} + n_3\sqrt{m_3}}{n_1\eta_1\sqrt{m_1} + n_2\eta_2\sqrt{m_2} + n_3\eta_3\sqrt{m_3}} \tag{30}$$

Below are the linear solutions for the case where J_2=-J_1 and J_3=0. The solutions have an exponential as well as quadratic terms. The ternary case is solved in the same manner as the binary case. The results, in Fig. 4 and again for i=2.0 A/cm^2 in Figs. 5 and 6, are similar. Here n_3 is the diluent, which we choose to be CO_2. This case differs from the binary case in that the metered gases are H_2 and CO_2, one of the gas pairs used by Jiang and Virkar.[9]

$$n_1(x) = n_{10} - \frac{D_{23}n_{30}}{D_{13} - D_{23}}\left(1 - \frac{D_{13}}{D_{12}}\right)\left(1 - \text{Exp}\left[\left(\frac{1}{D_{13}} - \frac{1}{D_{23}}\right)J_1 x\right]\right)$$
$$- J_1\left(\frac{1}{D_1} + \frac{n_{10} + n_{20} + n_{30}}{D_{12}}\right) x + \frac{J_1^2}{2D_{12}}\left(\frac{1}{D_1} - \frac{1}{D_2}\right) x^2 \tag{31}$$

$$n_2(x) = n_{20} + \frac{D_{13}n_{30}}{D_{13} - D_{23}}\left(1 - \frac{D_{23}}{D_{12}}\right)\left(1 - \text{Exp}\left[\left(\frac{1}{D_{13}} - \frac{1}{D_{23}}\right)J_1 x\right]\right)$$
$$+ J_1\left(\frac{1}{D_2} + \frac{n_{10} + n_{20} + n_{30}}{D_{12}}\right) x - \frac{J_1^2}{2D_{12}}\left(\frac{1}{D_1} - \frac{1}{D_2}\right) x^2 \tag{32}$$

$$n_3(x) = n_{30} \operatorname{Exp}\left[\left(\frac{1}{D_{13}} - \frac{1}{D_{23}}\right) J_1 x\right] \tag{33}$$

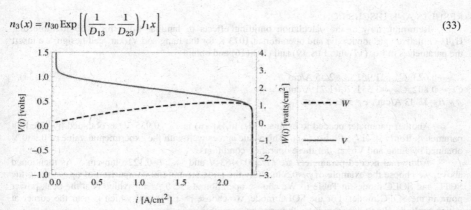

Figure 4. Total fuel cell voltage as a function of current density for anode thickness = 0.11 cm. The current cannot go negative because exhaust is not being pumped in.

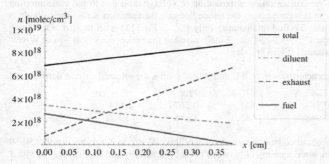

Figure 5. Concentration as a function of distance x along a typical anode pore path.

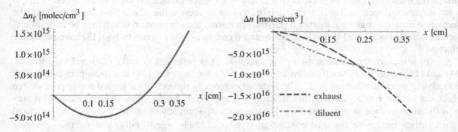

Figure 6. The difference between the solutions with and without laminar flow for the ternary case.

RESULTS AND DISCUSSION

Returning now to the calculation omitting effects of laminar flow, for a 50/50 molecular H_2/H_2O metered gas input ratio and operation at 1073 K for the Jiang and Virkar[9] cell design, we insert the parameters in Eq. (17) into Eqs. (9) and (14)-(16) and obtain

$$C_{aAP}=0.623, \; C_{aCP}=1.961, \; A_1=2.55 \text{ A/cm}^2, \qquad\qquad (34)$$
$$C_{bAP}=0.802, \; C_{bCP}=3.551, \; B_1=1.71 \text{ A/cm}^2,$$
$$A_2=B_2=13.43 \text{ A/cm}^2, \; c_0=5.03 \text{ A/cm}^2.$$

Another parameter needed to evaluate $V(i)$ in Eq. (6) is $V_0=0.935$ V, as obtained by inserting parameters from Eq. (17) into Eq. (11). This value agrees well with the experimental value of 0.900 V obtained by Jiang and Virkar[9] for these operating conditions.

Additional needed parameters are $kT/q=0.0463$ V and $V_{ohm}/i=0.122$ ohm-cm^2. As mentioned above, we choose the example of $p_f=0.5$ to allow the most meaningful comparison of operation in the SOFC and SOEC modes in Table I. We choose operation at $i=2.0$ A/cm^2 which is at the peak power point in the SOFC mode. For the SOEC mode, we choose $i=-1.5$ A/cm^2 which is near the corner at which applied voltage rises rapidly with increased magnitude of i.

In Table I, for the above parameters we list V_0-V_a which takes only the activation polarization V_a into account and which is found by setting C_{aCP}, C_{bCP}, A_2, B_2, and V_{ohm} to zero in Eqs. (13)-(16). We also list $V_0-V_a-V_c$, where V_c is the concentration polarization or voltage drop due to gas concentration variation across the electrodes, in this case across the anode because the variation across the cathode is considered small and is neglected. To find this voltage, only V_{ohm} in Eq. (13) is set to zero. Of course, we must list the terminal voltage $V=V_0-V_a-V_c-V_{ohm}$. Finally, we list the power density $W=Vi$ in W/cm^2 produced (SOFC mode) or consumed (SOEC mode).

Table I. Voltage and power characteristics in SOFC and SOEC modes for selected current densities.

Mode	i, A/cm^2	V_0-V_a, V	$V_0-V_a-V_c$, V	V, V	W, W/cm^2
SOFC	2.0	0.888	0.811	0.567	1.134
SOEC	-1.5	0.968	1.065	1.248	-1.872

We note that Jiang and Virkar[9] obtained peak power density of 0.95 W/cm^2 at 2.0 A/cm^2 for this 50/50 H_2/H_2O input gas mixture. The good agreement between model predictions and experiment[9] in Fig. 1 and Table I is gratifying.

Now we will examine Eqs. (14), (15), (34), and Table I to discuss the effect of changing various parameters in order to improve SOFC and SOEC performance for this particular cell. First, it is no surprise as seen in Table I that it is important to reduce the ohmic loss, by improving the electrolyte material, making it thinner if possible, or raising the temperature which may be undesirable. Second, we see from Eq. (34) that because A_2 and B_2 are large, the i^2 terms in Eq. (13) have negligible effect on V.

To minimize the concentration polarization V_c, it is sufficient to make C_{aCP} and C_{bCP} in Eqs. (14) and (15) respectively small compared to unity. We see from Eq. (34) that these parameters need reduction to meet this desired goal. The tortuosity comes in squared, so a graded anode pore design which can reduce τ is desirable. The thickness w is important, but for an anode-supported cell it may be difficult to reduce w substantially. The porosity ϕ is already 0.54, so it cannot be increased significantly. Increasing the total concentration n_i would help significantly, also for the activation polarization, and this can be accomplished by increasing the operating pressure by a few bars. For larger increase in pressure, Eqs. (14) and (15) show that there are diminishing effects in increasing pressure, as we noted previously,[4] unless the input gas flow rate is increased accordingly. Others have pointed out that increasing power output by increasing pressure may result in excessive heating. Increasing the input gas flow rate to increase i_l will not help substantially because i_l appears not only in the numerator, but also in the C terms in the denominator. Of course, performance can be increased in the SOFC mode by increasing p_f, and in the SOEC mode by decreasing p_f. Finally, D_{Kf} and D_{Ke} can be increased by increasing the pore diameter, but this is undesirable because it will reduce the triple phase boundary parameter f that appears in the activation polarization constants C_{aAP} and C_{bAP} which we will now discuss.

One might think that because f is so small compared to unity, that its increase would be of great help. However, C_{aAP} and C_{bAP} are already near unity or smaller, so it seems more important not to make any design changes that reduce f. The only other parameters in the activation polarization constants that we have some control over, are the fuel and exhaust gas molecular weights. We see that the low mass of H_2 is helpful in C_{aAP}, and it turns out to be helpful also in C_{aCP} because molecular mass enters into the square root power in the denominator of D_{Kf}.

Another check on the validity of our model, besides finding that it gives quite good agreement with experiment[9] for an H_2/H_2O input gas mixture, is to see whether it predicts the much weaker power output for CO fuel compared to H_2 fuel when all other operating parameters are the same. In particular, we can look at the maximum "cutoff" current attainable in the SOFC mode, given by A_1 in our model. The only new parameters needed for CO and the exhaust gas CO_2 for this calculation, compared to those for H_2 and H_2O in Eq. (17), are D_{Kf}=3.00 cm^2/s, D_{efl}=1.41 cm^2/s, and m_f=4.65x10^{-26} kg. The results of this comparison appear in Table II.

Table II. Comparison of cutoff currents for 50/50 CO/CO_2 and H_2/H_2O anode gas inputs.

Fuel/Exhaust	C_{aAP}	C_{aCP}	A_1=i_{cutoff}, A/cm^2	i_{cutoff}, meas. (Ref. 9)*
CO/CO$_2$	2.333	9.367	0.720	0.88
H$_2$/H$_2$O	0.623	1.961	2.552	3.2

*Based on short extrapolations of Jiang and Virkar data.

The most important result is that our model predicts almost exactly the 3.6 ratio of cutoff current for H_2 fuel compared to CO fuel. This large ratio has previously been attributed to Ni in the anode being a much better catalyst for H_2 than for CO. Our model does not consider catalysis. Instead, this large ratio comes from three sources, namely the much higher impingement speed of the much lighter H_2 molecules, and the much higher Knudsen (D_{Kf}) and intermolecular collision (D_{efl}) diffusion constants for H_2 molecule collisions with the pore walls and with the H_2O exhaust gas molecules, respectively. Besides predicting this cutoff current ratio correctly, the model comes close to predicting the correct magnitudes for both cutoff currents.

CONCLUSIONS

First, we discuss the strong and weak points of this model for SOFC/SOEC electrode gas flow and $V(i)$ characteristics.

A strong point is that it reproduces both the open-circuit voltage V_0 and the measured[9] $V(i)$ curves for a particular SOFC quite well, even though the model has only one adjustable parameter, namely υ which is taken as being the same for the anode and cathode interfaces. Also, it accounts for the lower output power for CO vs. H_2 fuel when all other operating conditions are the same, without making the usual assumption of Ni in the anode being a better catalyst for H_2 than for CO. It explains the "law of diminishing returns" for power increase as a function of operating pressure. As for gas flow, it shows that even for equal and opposite fuel and exhaust molecular flow in the anode passages, the increase in total pressure across the anode should not be neglected.

A weak point is that only one reaction mechanism, namely direct gas impact and reaction at the TPB sites, is assumed. No separate dissociation or surface diffusion steps are considered. No reaction barrier over and above the net energy change is assumed. The assumptions concerning ν and $1-\nu$ factors in the attempt current densities have no experimental verification. The agreement with experiment despite neglect of these matters does not prove our model is correct; it merely indicates that it could be correct.

We find that the use of H_2 as fuel is helpful for SOFC performance because of its low mass and corresponding higher velocity toward the reaction sites, in addition to being very desirable environmentally because the only exhaust product is steam.

We also find that the Jiang and Virkar[9] SOFC design has some room for improvements, but most of these are at the cost of undesirable side effects, so their design is not far from optimum for H_2 fuel. For CO and presumably other fuels of higher molecular weight, the design is farther from optimum. As one goes to heavier fuel molecules, there is more incentive to go to electrolyte-supported cells rather than anode-supported cells. Our model indicates that cathode-supported cells will have O_2 flow limitations unless pure oxygen is used.

Finally, we have shown that for practical SOFC/SOEC designs, consideration of laminar flow has a small effect on the analysis results because of the need for small diameter electrode pores in order to have sufficient triple phase boundary length.

ACKNOWLEDGEMENT

This work was supported by DOE under subcontract DE-AC06-76RL01839 from Battelle Memorial Institute and PNNL.

REFERENCES
[1] V. H. Schmidt, C.-L. Tsai, and L. M. Lediaev, Determination of Anode-Pore Tortuosity from Gas and Current Flow Rates in SOFC's, *Advances in Solid Oxide Fuel Cells III: Ceramic Engineering and Science Proceedings*, **28** (4), 129-140 (2007).
[2] V. H. Schmidt and C.-L. Tsai, Anode-Pore Tortuosity in Solid Oxide Fuel Cells Found from Gas and Current Flow Rates, *J. Power Sources*, **180**, 253-264 (2008).
[3] V. H. Schmidt, Dynamic First-Principles Molecular-Scale Model for Solid Oxide Fuel Cells, *ECS Transactions, Design of Electrode Structures*, **6** (21), 11-24 (2008).
[4] V. H. Schmidt and L. M. Lediaev, Pressure and Gas Concentration Effects on Voltage vs. Current Characteristics of a Solid State Fuel Cell and Electrolyzer, *Proceedings of the 32nd International Conference and Exposition on Advanced Ceramics and Composites, Daytona Beach, Florida, January 27-February 1, 2008.*
[5] J. Bear, *Dynamics of Fluids in Porous Media*, Dover Publ., Inc., New York (1988).
[6] M. Ni, M. K. H. Leung, and D. Y. C. Leung, A Modeling Study on Concentration Overpotentials of a Reversible Solid Oxide Fuel Cell, *J. Power Sources*, **163**, 460–466 (2006)
[7] Y. S. Touloukian (Ed.), *Thermophysical Properties of Matter, The TPRC Data Series—Viscosity*, Vol. **11**, 18a, TPRC, New York (1975).
[8] B. Todd and J. B. Young, Thermodynamic and Transport Properties of Gases for Use in Solid Oxide Fuel Cell Modeling, *J. Power Sources*, **110**, 186-200 (2002).
[9] Y. Jiang and A. V. Virkar, Fuel Composition and Diluent Effect on Gas Transport and Performance of Anode-Supported SOFCs, *J. Electrochem. Soc.*, **150** (7), A942-A951 (2003).
[10] R. Suwanwarangkul, E. Croiset, M. W. Fowler, P. L. Douglas, E. Entchev, and M. A. Douglas, Performance Comparison of Fick's, Dusty-Gas and Stefan-Maxwell Models to Predict the Concentration Overpotential of a SOFC Anode, *J. Power Sources*, **122**, 15 July 2003, 9-18 (2003).
[11] W. Lehnert, J. Meusinger, and F. Thom, Modeling of Gas Transport Phenomena in SOFC Anodes, *J. Power Sources*, **87**, 57-63 (2000).

[12] R.S. Gemmen and J. Trembly, On the Mechanisms and Behavior of Coal Syngas Transport and Reaction within the Anode of a Solid Oxide Fuel Cell, *J. Power Sources*, **161**, 1084-1095 (2006).

[13] M.M. Hussain, X. Li, and I. Dincer, Mathematical Modeling of Transport Phenomena in Porous SOFC Anodes, *Internat. J. Thermal Sciences*, **46**, 48-56 (2007).

Oxide/Proton Conductors

TEMPERATURE AND PREASSURE ASSISTED CUBIC TO RHOMBOHEDRAL PHASE TRANSITION IN $Sc_{0.1}Ce_{0.01}ZrO_2$ BY MICRO-RAMAN

Svetlana Lukich, Cassandra Carpenter, Nina Orlovskaya
Department of Mechanical, Material and Aerospace Engineering, University of Central Florida
Orlando, FL, USA

ABSTRACT

It was recently shown that the high temperature cubic $Sc_{0.1}Ce_{0.01}ZrO_2$ is metastable at room temperature and could be easily transformed to the thermodynamically stable rhombohedral phase upon annealing at $350^{\circ}C$-$400^{\circ}C$ after a certain period of time. Such transition, though not drastically important for operation of Intermediate Temperature Solid Oxide Fuel Cell, still might degrade the performance of $Sc_{0.1}Ce_{0.01}ZrO_2$ electrolyte upon thermal cycling. Therefore, the properties of the cubic and rhombohedral $Sc_{0.1}Ce_{0.01}ZrO_2$ phases have to be studied.

Here, we report the vibration response of cubic and rhombohedral $Sc_{0.1}Ce_{0.01}ZrO_2$ studied both at room and high-temperatures. Stress and temperature-assisted phase transition from cubic to rhombohedral phase was detected during in-situ Raman spectroscopy experiments. The conditions which promote either the formation of rhombohedral phase or retention of cubic phase upon temperature or stress applications have been determined.

INTRODUCTION

Sc_2O_3 doped ZrO_2 ($ScZrO_2$) ceramics have recently attracted a significant interest as a novel promising electrolyte material for lower temperature SOFCs due to its excellent ionic conductivity [1,2,3]. There have been numerous reports on the high ionic conductivity of $ScZrO_2$ ceramics [4, 5] which was reported to be near twice as high as other ZrO_2 based electrolytes [6]. Most of the studies of $ScZrO_2$ ceramics were performed on the materials with 8-12 mol% doping level of Sc_2O_3, where a cubic phase is a main single phase at 700-$800^{\circ}C$ operating temperatures. The drawback of $ScZrO_2$ has been also reported as an ordering of vacancies over time, called the aging phenomenon, accompanied by a phase transition to a lower symmetry rhombohedral phase, resulting in decreased conductivity [6]. The highly conductive cubic phase is not stable below $650^{\circ}C$ causing the abrupt decrease in ionic conductivity during cooling in the $ScZrO_2$ [7, 8]. It is known that in 11 mol% Sc_2O_3 – 89 mol% ZrO_2, a cubic to rhombohedral phase transition occurs when the temperature decrease below $600^{\circ}C$ [9, 10]. It was reported [11, 12] that when ZrO_2 is stabilized with a small amount of CeO_2 along with Sc_2O_3, it no longer exhibits an unfavorable phase transition, making this material a very promising option for intermediate temperature electrolytes.

In [11], the commercially available 10 mol% Sc_2O_3 – 1mol% CeO_2 – ZrO_2 ($Sc_{0.1}Ce_{0.01}ZrO_2$) manufactured by Daiichi Kigenso Kagaku Kogyo (DKKK, Japan) has been reported to have a stable cubic phase, superior electrical properties and excellent high temperature long term operating characteristics of single cells using $Sc_{0.1}Ce_{0.01}ZrO_2$ as an electrolyte material. Contrary, the reversible very slow cubic to rhombohedral and rhombohedral to cubic phase transitions at 300-$500^{\circ}C$ has been reported upon heating of $Sc_{0.1}Ce_{0.01}ZrO_2$ ceramics [13], which were probably overlooked in other studies due to extremely slow kinetics of cubic to rhombohedral phase transition upon heating. However, it is not expected that these transitions could have a significant effect on $Sc_{0.1}Ce_{0.01}ZrO_2$ electrolyte performance since they occur at lower temperatures and could simply be passed by during heating up or cooling down cycles of the cells. It was also reported that the kinetics of the phase transition is a strong function of the grain size of the $Sc_{0.1}Ce_{0.01}ZrO_2$ ceramics [13], therefore the cubic to β transition could be avoided if the grain size of $Sc_{0.1}Ce_{0.01}ZrO_2$ ceramics falls below a certain critical limit. It was also found that the coefficient of thermal expansion of cubic $Sc_{0.1}Ce_{0.01}ZrO_2$ is very close to the Y_2O_3 stabilized ZrO_2 (YSZ) which is a good indicator that $Sc_{0.1}Ce_{0.01}ZrO_2$ ceramics are a perfect candidate for substitution of YSZ electrolyte for IT SOFCs.

While XRD patterns are determined by the arrangements of the cations (Zr, Sc, Ce) in the fluorite lattice, laser-excited Raman spectra are sensitive to cation-oxygen bands and can easily give invaluable information about the local distortions around cations as well as can identify the disorder in the oxygen sublattice. The different ZrO_2 structures (monoclinic, tetragonal, rhombohedral, and cubic) all have characteristic signatures in their spectra, which enable them to be easily distinguish and even quantify their amount. Therefore, the goal of this paper is to study the cubic and rhombohedral $Sc_{0.1}Ce_{0.01}ZrO_2$ ceramics using Raman spectroscopy, which is a very powerful technique to reveal vibrational properties of metal-oxygen bonds as well as temperature and stress induced deformation and phase transitions in zirconia.

EXPERIMENTAL DETAILS

The 10mol% Sc_2O_2 – 1mol% CeO_2 – ZrO_2 ($Sc_{0.1}Ce_{0.01}ZrO_2$) powder produced by Daiichi Kigenso Kajaku Koguo (DKKK, Japan) has been sintered at 1500°C for 2 hours with a 10°C/min heating/cooling rate in air to almost full density. The XRD confirms that the material consist of the cubic phase upon cooling after sintering. The cubic $Sc_{0.1}Ce_{0.01}ZrO_2$ samples were grinded and polished in order to obtain the mirror surface and then they were thermally etched at 1300°C for 1 hour to reveal the grain boundaries. A portion of the thermally etched cubic $Sc_{0.1}Ce_{0.01}ZrO_2$ samples were annealed at 375°C for 12 hours in air in order to convert them to the β rhombohedral phase, which is stable in 25 - 400°C temperature range [13]. Both cubic and rhombohedral $Sc_{0.1}Ce_{0.01}ZrO_2$ phases were indented using Vickers hardness tester (LECO M-400) with a load of 9.8N. The hardness and fracture toughness of the ceramics have been calculated using the length of the impression diagonals and cracks originating from the corners of the impressions, respectively. Optical micrographs were taken using an Olympus confocal microscope (LEXT OLS3000-IR). The processing of the Sc_2O_3 doped ZrO_2 ($ScZrO_2$) ceramics as well as selected properties, such as hardness, fracture toughness, Young's modulus, are presented in details elsewhere [13,14,15].

Renishaw InVia Raman microscope was used to study the vibrational spectra of $Sc_{0.1}Ce_{0.01}ZrO_2$ ceramics. The Raman microscope system comprises a laser (532 nm line of solid Si or near infrared 785nm) to excite the sample, a single spectrograph fitted with holographic notch filters, and an optical microscope (a Leica microscope with a motorized XYZ stage) rigidly mounted and optically coupled to the spectrograph. The generated laser power was 25 mW. Before collecting spectra of $Sc_{0.1}Ce_{0.01}ZrO_2$ the spectrometer was calibrated with a Si standard using a Si band position at 520.3 cm^{-1}. The average collection time for a single spectrum was 20s. High temperature Raman spectroscopy was performed using a TMS 600 and TMS 1500 heating stage (Linkam Scientific Instruments Ltd, UK) by heating/cooling of the samples to/from 400°C and 1000°C, respectively. For the high temperature experiment the incident and scattered beams were focused with a long working distance 50x objective, which maintained a laser spot as small as 2-3 μm. A 10°C/min heating/cooling rate was used for high temperature experiments. Room temperature Raman spectra were collected from different points of interest on the sample surface, such as on the polished surface at different locations as well as inside or outside the Vickers impressions. For room temperature and area mapping experiments the short working distance 100x objective was used. To produce two dimensional (2D) maps, Renishaw Wire 2.0 software with a mixed Lorentzian and Gaussian peak fitting function was used. The system's peak fitting results were plotted to create a position map with a spectral resolution better than 0.2 cm^{-1}. The total time of spectrum collection was decreased to 3 s per point in the case of 2D mapping and the total acquisition time to collect all spectra for one map never exceeded 24 hours.

RESULTS AND DISCUSSION

The whole spectral range of cubic and β-$Sc_{0.1}Ce_{0.01}ZrO_2$ phases collected using both 532nm and 785nm lasers are presented in Fig. 1. While the spectra collected using 532nm Si laser shows the most prominent features at ranges $100 - 1000cm^{-1}$ and $7200 - 8000cm^{-1}$ both for cubic and rhombohedral structures, the spectra collected using NIR 785nm laser shows the strongest bands in the $1000 - 2000cm^{-1}$ range. The bands with a Stokes shift higher than $800cm^{-1}$ from the laser excitation line have been previously observed in zirconia based oxides [16, 17, 18], and bands have been assigned to electronic transitions in impurity ions [19] or to phonon–mediated de–excitation of excited states of the impurity–doped ZrO_2 lattice [20]. As one can see from Fig. 1, a number of peaks showing up both in cubic and β-$Sc_{0.1}Ce_{0.01}ZrO_2$ in 1000-$2000cm^{-1}$ range using 785nm excitation, are completely missing when 532nm laser is used. While the exact nature of the bands is not obvious, they could be tentatively assigned to the appearance of the luminescence bands related to Ln^{3+} (Ln - Pr^{3+}, Nd^{3+}, Mo^{3+}, Er^{3+}) impurities ions [21], and more research needs to be performed to establish the origin of the 1000-$2000cm^{-1}$ bands. Obviously, it is useful to study the vibrational response of the $Sc_{0.1}Ce_{0.01}ZrO_2$ using at least two lasers, as the structural information from the analysis of Raman spectra measured with only one laser could be incorrectly interpreted.

In order to study the effect of stress on vibrational properties of cubic and β-$Sc_{0.1}Ce_{0.01}ZrO_2$, the polished surface was indented using Vickers diamond indenter. For comparison, the spectra taken both from the non-deformed surface and from the center of the Vickers impressions of cubic and β-$Sc_{0.1}Ce_{0.01}ZrO_2$ are shown. While, the position shift and change of the intensities and Full Width at Half Maximum (FWHM) of certain bands can be observed, the structures of cubic and β phases remain the same after the deformation by the sharp Vickers indenter. Thus, no phase transformation can be reported upon indentation, only the strained structures.

The spectra of cubic and β-$Sc_{0.1}Ce_{0.01}ZrO_2$ phases in $100 - 1000cm^{-1}$ range are presented in Fig. 2. The bonds located at Stokes shifts lower than $800cm^{-1}$ from the excitation line are assigned to Raman active lattice phonons [22, 23].The reported Raman spectrum of pure submicron cubic zirconia consists of a weak broad line assigned to a single allowed Raman mode F_{2g} symmetry [24]. Only oxygen atoms move in this mode, therefore the frequency should be independent of the cation mass,

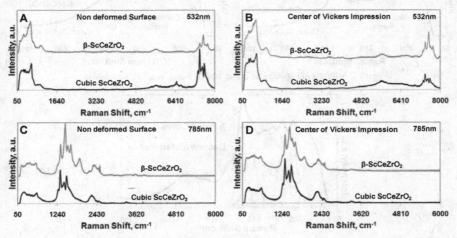

Figure 1. The Spectral Range of $Sc_{0.1}Ce_{0.01}ZrO_2$ Electrolyte Ceramics

and it was reported that in fluorite structures, CeO_2, ThO_2, UO_2 the mode has been observed near the $460 - 470cm^{-1}$ region [25, 26]. However, the spectrum of cubic $Sc_{0.1}Ce_{0.01}ZrO_2$ contains a number of bands. Some of them can be located at 239, 315, 378, 480, and $623cm^{-1}$ (Fig. 2). The similar spectrum of cubic $Sc_{0.1}Ce_{0.01}ZrO_2$ cubic electrolyte was published in [27]. The deconvolution of the peaks is rather problematic and can be ambiguous. Up to 10 separate peaks located below $800cm^{-1}$ have to be assigned in order to obtain a good match between experimental and calculated data   (Fig. 3). The appearance of the peaks, which are not allowed in the cubic fluorite structure, can be attributed to the disorder of ionic defects in the oxygen sublattice. Substitution of Zr^{4+} by Sc^{3+} results in the formation of high quantities of oxygen vacancies, and such high defect concentration can lead to a violation of the selection rules and allows the appearance of additional modes that are forbidden for the cubic fluorite structure [26].

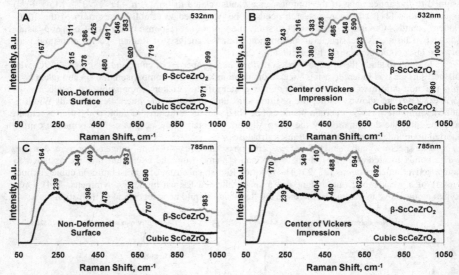

Figure 2. Vibrational spectra of cubic and β- $Sc_{0.1}Ce_{0.01}ZrO_2$ ceramics

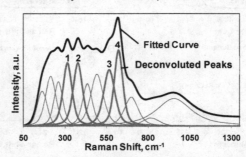

Figure 3. Deconvoluted peaks of $Sc_{0.1}Ce_{0.01}ZrO_2$ ceramics, 532nm Si laser

The differences of the cubic and rhombohedral $Sc_{0.1}Ce_{0.01}ZrO_2$ are clearly visible, especially when spectra are taken using 532nm Si laser. As it was already reported in [10] the Raman spectrum of β phase is rather complex, presented by a characteristic broad continuum with many small peaks rising above background.

The asymmetry of the $632cm^{-1}$ band on the left low frequency side is typical for disordered systems, as it was indicated by Kosacki [26]. The heating of the cubic $Sc_{0.1}Ce_{0.01}ZrO_2$ up to 1000°C did not remove the asymmetry (Fig. 4), however the intensities of the bands decreased and a significant broadening of the peaks has been observed upon temperature increase. Due to the broadening of the 315 and $378cm^{-1}$ peaks, marked as #1 and #2 at Fig. 4, the coalesced into one broad peak starting from 800°C, which was impossible to deconvolute into two peaks reliably.

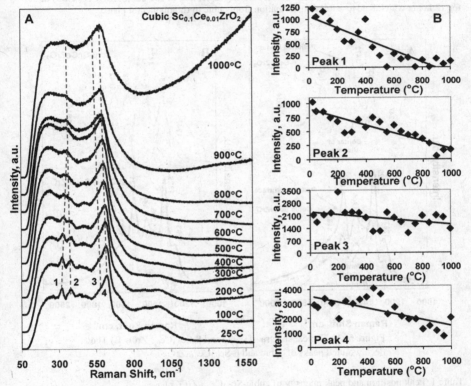

Figure 4. A) High temperature spectra of cubic of $Sc_{0.1}Ce_{0.01}ZrO_2$ ceramics heated up to 1000°C with 100°C step size and 10°C/min heating rate. A dwell time of 5 minutes was used at each temperature before collection. B) Peak intensity vs. temperature for peaks # 1, 2, 3, and 4 of cubic $Sc_{0.1}Ce_{0.01}ZrO_2$. 532nm Si laser

Since the spectra collected below 800cm^{-1} of both cubic and β- $Sc_{0.1}Ce_{0.01}ZrO_2$ are very complex (Fig. 2 and 3) they are not very easy to use for the mapping experiments, which can help to detect the phase transitions or quantify the residual stresses. Therefore, the $1000 - 2000cm^{-1}$ range of the spectrum collected by NIR 785nm laser has been chosen for the collection of 2D maps. Fig. 5A represents the typical spectra of cubic and β-phases. Spectra taken both from the polished surface and from the center of the Vickers impression do not show a significant difference. The deconvoluted peaks of cubic and β-$Sc_{0.1}Ce_{0.01}ZrO_2$ are shown in Fig. 5B. Their parameters, such as peak position and peak intensity, are presented in Table 1 and 2. Confocal optical micrographs of the cubic and β-$Sc_{0.1}Ce_{0.01}ZrO_2$ are shown in Fig. 6. While the cubic $Sc_{0.1}Ce_{0.01}ZrO_2$ has a smooth and flat surface, where grain boundaries are clearly visible, the β-$Sc_{0.1}Ce_{0.01}ZrO_2$ has formed surface termination steps. The formation of the β structure has occurred during annealing of the cubic $Sc_{0.1}Ce_{0.01}ZrO_2$ at 375°C for 12 hours which lead to a full transformation from cubic to rhombohedral phase.

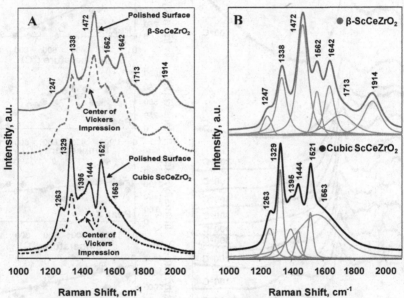

Figure 5. A) Typical spectra of cubic and β-$Sc_{0.1}Ce_{0.01}ZrO_2$; B) The deconvoluted peaks of cubic and β-$Sc_{0.1}Ce_{0.01}ZrO_2$

Table 1. Peak position and peak intensity of cubic-$Sc_{0.1}Ce_{0.01}ZrO_2$ phase

Cubic-ScCeZrO$_2$		Peak1	Peak2	Peak3	Peak4	Peak5	Peak6
Position	Non Deformed Surface	1,263	1,329	1,395	1,444	1,521	1,563
	Vickers Impression	1,261	1,332	1,398	1,446	1,528	1,571
Intensity	Non Deformed Surface	9,587	28,695	9,374	8,892	14,687	14,148
	Vickers Impression	5,109	14,677	5,260	5,462	7,164	8,376

Table 2. Peak position and peak intensity of β-$Sc_{0.1}Ce_{0.01}ZrO_2$ phase

β-ScCeZrO$_2$		Peak1	Peak2	Peak3	Peak4	Peak5	Peak6	Peak7
Position	Non Deformed Surface	1,247	1,338	1,472	1,562	1,642	1,713	1,914
	Vickers Impression	1,254	1,337	1,469	1,561	1,641	1,725	1,913
Intensity	Non Deformed Surface	2,162	9,593	18,206	7,261	8,202	3,397	5,624
	Vickers Impression	5,021	16,906	22,750	9,382	11,092	4,656	6,054

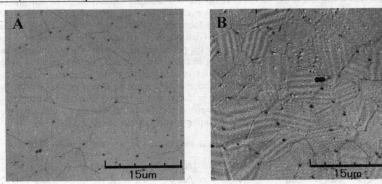

Figure 6. Confocal images of cubic (A) and β-ScCeZrO$_2$ (B)

The confocal optical micrographs of the Vickers impressions obtained by identifying cubic and β-$Sc_{0.1}Ce_{0.01}ZrO_2$ are shown in Fig. 7. A total of 20 impressions were made for each sample and with a 9.8N load. The impressions were used for the measurements of hardness and indentation fracture toughness. $H_v = 14.02 \pm 0.3$ GPa and $K_{1C} = 2.03 \pm 0.7$ MPa·m$^{1/2}$ are reported for cubic $Sc_{0.1}Ce_{0.01}ZrO_2$ and $H_v = 13.78 \pm 0.4$ GPa and $K_{1C} = 1.68 \pm 0.2$ MPa·m$^{1/2}$ are reported for β-$Sc_{0.1}Ce_{0.01}ZrO_2$.

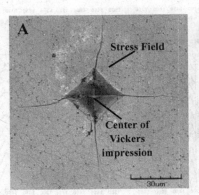 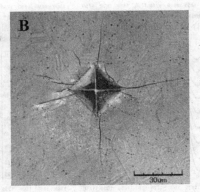

Figure 7. Confocal optical micrographs of impressions made by a Vickers indenter in A) cubic of $Sc_{0.1}Ce_{0.01}ZrO_2$ and B) β-$Sc_{0.1}Ce_{0.01}ZrO_2$ ceramics used to determine hardness and indentation fracture toughness

In order to determine the stability of the cubic and β-phases in 25 - 400°C temperature range the heating/cooling experiments were performed where spectra were collected from three different locations on the sample's surface during heating and cooling. Three different samples of cubic $Sc_{0.1}Ce_{0.01}ZrO_2$ and three different samples of β-$Sc_{0.1}Ce_{0.01}ZrO_2$ were indented using a Vickers indenter at 9.8N. For each indented sample, a separate location has been chosen: a) a non-deformed polished surface far away from the impression; b) the center of the Vickers impression; c) a location close to the Vickers impression where a stress should develop due to a deformation introduced by indentation. The last two locations, a center of the Vickers impression and a stress field point, are shown in Fig. 7A.

The spectra collected form three different locations in cubic and β-phases during heating and cooling up to 400°C are presented in Fig. 8. It was detected (Fig. 8A) that while cubic phase is retained as cubic on the polished non-deformed surface for the performed heating experiment, it would partially transform to a rhombohedral phase upon cooling in the center of the Vickers impression, where the mixture of cubic and rhombohedral phases could be detected at room temperature after cooling. However at the stress field deformation zone location a full transformation from the cubic to β-phase would occur, thus only the β-phase spectrum can be found and no cubic phase can be detected. At the same time, it was found that the rhombohedral β-phase is stable in the whole 25 - 400°C temperature range at all three locations both upon heating and upon cooling conditions (Fig. 8B).

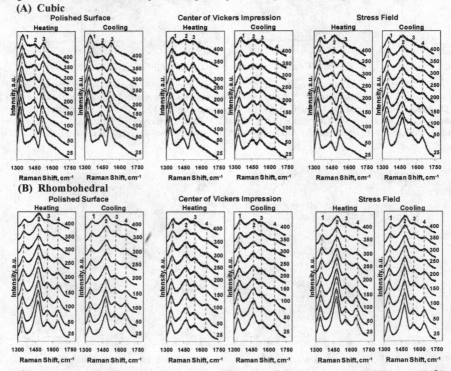

Figure 8. Spectra of cubic (A) and rhombohedral (B) phase during heating and cooling up to 400°C

Interesting results were obtained when the position of the peaks of cubic and β phases were analyzed as a function of the locations and temperature (Fig. 9C). While the positions of the peaks of the cubic phase collected from the center of the Vickers impression were shifted to the higher wavenumbers relative to the peak positions collected from the polished, non-deformed surface and stress field (Fig. 9A), the opposite tendency was detected to occur for some of the peaks of the β phase (Fig. 9B). Thus, peaks #2 and #3 were shifted to the lower wavenumbers by ~10–15cm^{-1} relative to their positions at the polished and stress field surfaces. However, peak #1 hardly showed any shift in position. This difference can be clearly seen when a 2D map of the peak positions were created with Vickers impression made in cubic and β phases (Fig. 10). The maps were collected using ~1530cm^{-1} cubic and ~1465cm^{-1} rhombohedral peak positions as a function of location on the samples' surface. The high residual compressive stresses created inside the impression upon indentation caused the shift of the ~1530cm^{-1} peak to the higher wavenumbers in the cubic phase with the maximum ~15cm^{-1} shift observed in the center of the Vickers impression. On the opposite, the ~1465cm^{-1} peak position was shifted to the lower wavenumber upon indentation in the β-phase. The maximum ~10cm^{-1} shift was observed with a change from ~1476cm^{-1} for the center of the impression.

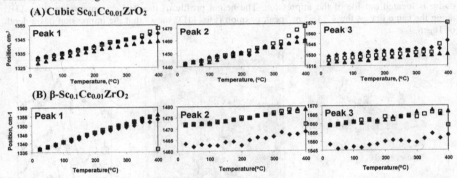

Figure 9. Position of the peaks of cubic (A) and β-phases (B) vs. temperature upon heating experiments; ▲ -Polished surface, ◆ - Center of Vickers impression, ☐ - Stress field

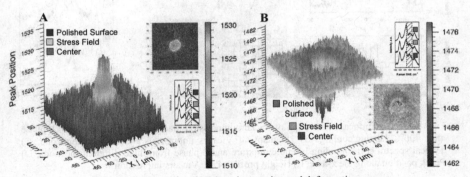

Figure 10. 3D maps of Vickers impression and deformation zone

After heating/cooling experiments (Fig. 8) of cubic phase in 20-400°C temperature range, it became clear that not only temperature, but also stress affects the phase composition of $Sc_{0.1}Ce_{0.01}ZrO_2$. Fig. 11A shows the confocal optical micrograph of the impression made in the cubic phase before heating and then heated/cooled to 400°C. The deformation zones formed around impression are clearly visible. This impression was used for mapping, and two of the maps showing formation of the rhombohedral β-phase in the stressed fields around the impression are presented in Fig. 11B. The first impression map includes the larger area of 180x180 μm, and the second map includes the more detailed view with a smaller 120x120 μm area of the same impression. The typical spectra, corresponding to cubic, β and a mixture of the phases, collected from the different locations on the maps are given in Fig. 11C. As one can see from the map, the cubic phase was mostly retained both further away outside of the deformation zone at the polished surface and inside of the Vickers impression after heating/cooling, however the cubic to β phase transformation has occurred inside of the stress field. The 1D line map of the ~1460cm^{-1} peak position along the x axis at 0 interceptions with the y axis of the map in shown in Fig. 11D. It shows that the peak position of the cubic phase inside the Vickers impression has shifted to the higher wave numbers relative to the non-deformed phase is located outside of the impression. The height profile of the Vickers impression (Fig. 11E) along the same line as the ~ 1460cm^{-1} peak position (Fig. 11D) shows that the impression has a depth of 10μm.

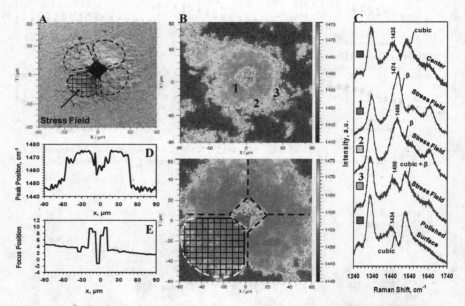

Figure 11. A) Confocal micrograph of Vickers impression made in cubic $Sc_{0.1}Ce_{0.01}ZrO_2$ phase after 400°C heating/cooling; B) 2D maps of Vickers impression and deformation zone; C) Typical spectra corresponding to cubic, mixture and β-phase from three different locations; D) Peack position along the x-axis; E) Height profile of the Vickers impression

CONCLUSIONS

The research on the vibrational behavior of cubic and rhombohedral $Sc_{0.1}Ce_{0.01}ZrO_2$ has shown that micro-Raman spectroscopy is a powerful tool which can be successfully used to characterize the fluorite structures and their stress states. The spectra of cubic and β- $Sc_{0.1}Ce_{0.01}ZrO_2$ have been collected using 532nm and 785nm lasers. While some of the bands, mostly in 100-800cm^{-1} range, were detected at the same positions by two lasers, some of the bands can be excited only using 785nm laser. The origin of these bands is not clear, however tentatively they can be assigned to the appearance of the luminescence bands related to Ln^{3+} impurity ions.

The in-situ heating/cooling experiments revealed that the cubic $Sc_{0.1}Ce_{0.01}ZrO_2$ is stable upon heating up to 1000°C, but it can transform to rhombohedral phase upon cooling if located near the Vickers impression. The rhombohedral β-$Sc_{0.1}Ce_{0.01}ZrO_2$ is stable upon heating in 25-400°C temperature range.

The bands of cubic and β-$Sc_{0.1}Ce_{0.01}ZrO_2$ located in 1000-2000cm^{-1} range have been selected to perform mapping experiments to characterize stress state in the material introduced by Vickers indentation as well as detect any possible phase transitions. The stress related shift of peak position has been detected both in cubic and rhombohedral β-$Sc_{0.1}Ce_{0.01}ZrO_2$ upon indentation. The peaks were shifted to the higher wavenumbers in the center of the Vickers impression in the cubic phase due to residual compressive stress after removal of the load, however most of the peaks (#2, 3 and 4) of the β-$Sc_{0.1}Ce_{0.01}ZrO_2$ were shifted to the lower wavenumbers in the center of the impressions. The further calibration experiments of the stress induced peak shift are required in order to quantify the residual stress in the $Sc_{0.1}Ce_{0.01}ZrO_2$. Both cubic and rhombohedral phases remain stable upon indentation and no phase transformation was detected for the experiments performed at room temperature. However, the cubic to β phase transformation has been detected when a cubic $Sc_{0.1}Ce_{0.01}ZrO_2$ sample with Vickers impression was heated to 400°C and then cooled down to room temperature. The visible structural changes have been detected in the deformation zones located around Vickers impressions after heating. The mapping experiments revealed that the β-phase has been formed in the deformation zones, however the cubic phase has been retained inside the impression and at the polished surface which was not subject to the deformation by contact loading.

ACKNOWLEDGMENT

This work was supported by the NSF DMR project number 0502765.

REFERENCES

[1] Z. Ze, Q. Zhu, Low Temperature Processing of Dense Nanocrystalline Scandia – Doped Zirconia (ScSZ) Ceramics, *Solid State Ionics*, 176, 37-38, 2791-2797 (2005).

[2] I. Kosacki, H. U. Anderson, Y. Mizutani, K. Ukai, Nonstoichiometry and Electrical Transport in Sc – doped Zirconia, *Solid State Ionics*, 152-153, 431-438 (2002).

[3] V. Kharton, F. M. B. Marques, A. Atkinson, Transport Properties of Solid Oxide Electrolyte Ceramics: A Brief Review, *Solid State Ionics*, 174, 1-4, 135-149 (2003).

[4] S.P.S. Badwal, S. F. Ciacehi, S. Rajendran, J. Drennan, An Investigation of Conductivity, Microstracture, and Stability of Electrolyte Compositions in the System 9mol% (Sc_2O_3-Y_2O_3) – ZrO_2 (Al_2O_3), *Solid State Ionics*, 109, 3-4, 167-186 (1998).

[5] O. Yamamoto, Y. Arachi, Y. Takeda, N. Jmaniski, Y. Mizutani, M. Kawai, Y. Nakanmra, Electrical Conductivity of Stabilized Zirconia with Yttria and Scandia, *Solid State Ionics*, 79, 127-142 (1995).

[6] Y. Arachi, Y. M. Sakai, Y. Yamamoto, Y. Takeda, N. Imaniski, Electrical Conductivity of the ZrO_2 – Ln_2O_3 (Ln- lanthanides) system, *Solid State Ionics*, 121, 1 – 4, 133 – 139 (1999).

[7] H. Yananura, N. Utsunomiya, T. Mari, T. Ateka, Electrical Conductivity in the System ZrO_2 – Y_2O_3 – Sc_2O_3, *Solid State Ionics*, 107, 3-4, 185-189 (1998).

[8]V. Arachi, T. Asai, O. Yamamoto, V. Takeda, N. Imaniski, K. Kawate, C. Tamakoshi, Electrical Conductivity of ZO_2 – Sc_2O_3 Doped with HfO_2, CeO_2 and Ga_2O_3, *J. Electrochem. Soc*, 148, 5, A520-A523 (2001).

[9]M. Yashima, T. Kakihana, M. Yoshimura, Metastable Phase Diagrams in the Zirconia Containing Systems Utilized in Solid Oxide Fuel Cell Aplications, *Solid State Ionics*, 86-88, 2, 1131-1149 (1996).

[10]H Fuijimori, M. Yashima, M. Kakihana, M. Yoshimura, The β-Cubic PhaseTransition of Scandia Doped Zirconia Solid Solution: Calorimetry, X-ray Diffraction, and Raman Scattering, *J. Appl. Phys*, 91, 10, 6493-6498 (2002).

[11]D. S. Lee, W. S. Kim, S. H. Choi, J. Kim, H. W. Lee, J. H. Le, Characterization of ZrO_2 Co-doped with Sc_2O_3 and CeO_2 Electrolyte for the Application of Intermediate Temperature SOFCs, *Solid State Ionics*, 176, 9-10, 33-39 (2005).

[12]Z. Wang, N. Chang, Z. Bi, Y. Dong , H. Zhang, J. Zhang, Z. Feng, C. Li, Structure and Impedance of ZrO_2 Doped with Sc_2O_3 and CeO_2, *Mater. Lett.* 59, 19-20, 2579-2582 (2005).

[13]S. Yarmolenko, J. Sanker, N. Bernier, N. Klimov, J. Kapat, N. Orlovskaya, Phase Stability and Sintering Behavior of 10mol% Sc_3O_3 – 1mol% CeO_2-ZrO_2 Ceramics, *J. Fuell Cell Sci. Tech.*, accepted (2008).

[14]A. Zevalkink, A. Hunter, M. Swanson, C. Johnson, J. Kapat, N. Orlovskaya, Processing and Characterization of Sc_2O_3-CeO_2-ZrO_2Electroyte Based Intermediate Temperature Solid Oxide Fuel Cells, *Mater. Res. Soc, Symp. Proc.*, 972 (2007)

[15]N. Orlovskaya, S. Lukich, N. Lugovy, J. Kuebler, Mechanical Properties of $Sc_{0.1}Ce_{0.01}ZrO_2$ Electrolyte Ceramics, unpublished results.

[16] T. Otake, H. Yugami, H. Naito, K. Kawanmra, T. Kawada, J. Mizusaki, Ce^{3+} Concentration in ZrO_2–CeO_2–Y_2O_3 System Studied by Electronic Raman Scattering, *Solid State Ionics*, 135, 663 (2000).

[17] V. M. Orera, R. I. Merino, F. Pena, $Ce^{3+}\leftrightarrow Ce^{4+}$ Conversion in Ceria-Doped Zirconia Single Crystals Induced by Oxido-Reduction Treatments. *Solis State Ionics*, 72, 224 (1994).

[18] J. Kaspar, P. Fornasiero, G. Baldueci, R. Di Monte, N. Hickey, V. Sergo, Effect of ZrO_2 content on textural and structural properties of CeO_2-ZrO_2 solid solutions made by citrate complexation route, *Inorg. Chim. Acta*, 349, 217 (2003).

[19] N. Maczka, E. T. G. Lutz, H. J. Verbuk, k. Oskam, A. Meijerink, J. Hanuza, N. Stuivinga, Spectroscopic Studies of Dynamically Compacted Monoclinic ZrO_2, *J. Phys. Chem. Solids*, 60, 1909 (1999).

[20]I. N. Asher, B. Papanicolaou, E. Anastassakis, Laser Excited Luminescence Spectra of Zirconia, *J. Phys. Chem. Solids*, 37, 221 (1976).

[21]P. Fornasicro, A. Spghini, R. Di Monte, M. Bettinelli, J. Kaspar, A. Bigotto, V. Sergo, M. Graziani, Laser – Excited Luminescence of Trivalent Lanthanide Impurities and Local Structure in CeO_2 – ZrO_2 Mixed Oxides, *Chem. Mater.*, 16, 1938 – 1944, (2004).

[22] G. Vlaic, R. Di Monte, P. Fornasiero, E. Fonda, J. Kaspar, M. Graziani, Redox Property–Local Structure Relationships in the Rh-Loaded CeO_2-ZrO_2Mixed Oxides, *J. Catal.*, 182, 378 (1999).

[23] M. Yashima, H. Arashi, N. Kakihana, M. Yoshimura, Oxygen-Induced Structural change of the Tetragonal Phase Around the Tetragonal-Cubic Phase Boundary in ZrO_2-$YO_{1.5}$ Solid Solutions, *J. Am. Ceram. Soc.*, 77, 1067 (1994).

[24]C. M. Philippi, K. S. Mazgiyasni, Infrared and Raman Spectra of Zirconia Polymorphs, *J. Am. Ceram. Soc.*, 54, 254 (1971).

[25]A. Feinberg, C. H. Perry, Structural Disorder and Phase Transitions in ZrO_2-Y_2O_3 System, *J. Phys. Chem. Solids*, 42, 513 (1981).

[26]I. Kosacki, V. Petrovsky, H. Anderson, P. Colombon, Raman Spectroscopy of Nanocrystalline Ceria and Zirconia Thin Films, *J. Am. Ceram. Soc.*, 85, 2646 (2004).

[27]H. Kishimoto, N. Sakai, T. Horita, K. Yamaji, Y. P. Xiong, M. E. Brito, H. Yokokawa, Rapid Phase Transformation of Zirconia in the Ni-ScSZ Cermet Anode Under Reducing Condition, *Solid State Ionics*, 179, 2037-2041 (2008).

SYNTHESIS AND ACTIVITY OF COBALT-DOPED BARIUM CERIUM ZIRCONATE FOR CATALYSIS AND PROTON CONDUCTION

Aravind Suresh, Joysurya Basu, Nigel M. Sammes[*], C. Barry Carter and Benjamin. A. Wilhite
Department of Chemical, Materials and Bio-molecular Engineering (CMBE) and Connecticut Global Fuel Cell Center (CGFCC), University of Connecticut, Storrs, CT, 06269
[*] Metallurgical & Materials Engineering Department, Colorado School of Mines, 1500 Illinois St., Golden, CO, 80401

ABSTRACT

Reforming high energy-density liquid fuels in the presence of a catalyst is a convenient means of generating hydrogen for energy conversion in fuel cells, which is always accompanied with several other by-products detrimental to the cell and the environment. For this reason, it is of particular interest to couple the fuel reforming and the hydrogen purification step together. A promising strategy for accomplishing this end is to utilize mixed ionic/electronic conducting ceramics with catalytic functionality and proton conductivity. Cobalt-doped barium cerate and zirconate perovskites ($BaCe_{1-\alpha-\beta-\gamma}Zr_\beta Co_\gamma O_{3-x}$) display both of these properties. Oxalate chemistry has been used to produce Co-doped powders at low (near ambient) temperatures starting with the nitrate salts of the chosen metals as the precursors. The structure and chemistry of the powder has been monitored at each stage of the preparation using XRD and TEM. The as-precipitated powder is amorphous and crystalline barium cerate, barium zirconate, barium cobaltate are all formed initially after moderate temperature heat treatments. The desired barium cerium zirconate phase with homogeneous Co distribution forms as a result of reaction at high temperatures.

INTRODUCTION

Growing world-wide demands for energy and environmental concerns, coupled with dwindling resources, has driven the search for a universal, emission-free fuel derivable from a broad spectrum of fuel sources ranging from fossil (e.g., coal, natural gas) and bio-mass (e.g. biogas, ethanol, butanol) hydrocarbons. Hydrogen can be extracted from any hydrocarbon resource, thus providing a universal energy currency for the 21st century; it can subsequently be converted directly to electrical energy via fuel cells, thus removing the Carnot-cycle limit on combustion efficiency for greater resource conservation.

Oxidative reforming of high energy-density liquid fuels such as light alcohols (e.g., methanol, ethanol, butanol) provides a convenient route for hydrogen generation. By-products of gas-phase catalytic reaction of hydrocarbons with oxygen and/or steam include carbon monoxide and dioxide, methane and water. Both carbon monoxide and carbon dioxide have been shown to reduce the long-term efficiency of proton-exchange membrane fuel cells via competitive adsorption on anode catalyst sites [1, 2]; for this reason hydrogen purification of the reaction product mixture via permselective membranes is required.

Hydrogen purification using dense thin-film membranes of palladium and its alloys (e.g., Ag-Pd, Cu-Pd) have been widely investigated as a method of hydrogen purification [3]. Drawbacks to this technology include high material costs and film vulnerability to corrosion upon exposure to reforming chemistries and carbon monoxide by-product [4]. Corrosion issues associated with palladium-based membranes can be mitigated via incorporation of porous catalytic layers over-top the permselective film, such that reforming reaction and subsequent purification of reaction products are coupled in sequence via gas diffusion [5]. This strategy is limited by Pd film stability at temperatures above 500°C [6], which prevents reforming of several candidate fuels (e.g., methane, propane, butanol) requiring reforming temperatures in excess of 500°C [7].

An alternate strategy described here is to employ electro-ceramic membrane materials capable of hydrogen extraction from liquid hydrocarbons with greater chemical and thermal stability than palladium-based counterparts. This can be accomplished by designing materials with (i) catalytic activity for hydrogen generation, and (ii) mixed proton-electronic conductivity (MPEC) for hydrogen purification. Resulting materials are expected to be capable of intimately coupling surface reactivity with bulk transport of hydrogen via electrochemical routes. Additional improvements in overall reaction rates and selectivity can be developed using electronic (Faradaic enhancement) and ionic (Non-Faradaic enhancement) potentials by coupling catalytic and electrochemical phenomena within a single material. This new approach is of special interest, since these enhancements in fuel reforming rates and hydrogen yields would be beyond the classical benefits in performance associated with conventional membrane reactors owing to Le Chatelier's principle [5, 8].

Ceramics materials based on CeO_2 have been studied for several years [9, 10]. Barium cerate was chosen as the principal material since doped barium cerate perovskites have been widely reported to display substantial proton conductivity under hydrogen/water environments [11, 12]. Additionally, several authors have reported ceria-supported noble metal catalysts [13-15] and ceria-based perovskites [16] as promising fuel reforming catalysts. Partial substitution of cerium with Zr has been reported to improve chemical stability of barium cerates in the presence of carbon dioxide, with an offset in proton conductivity [17, 18]. Cobalt-based catalysts have been identified as highly selective to partial oxidation and steam reforming of methanol and ethanol by multiple authors [5, 19]. In addition to its catalytic contributions, the divalent and trivalent (+2/+3) nature of cobalt is expected to contribute to both proton and p-type semiconductor electronic conduction [20]. Cobalt-doped barium cerate/zirconate blends were thus examined as a candidate material for SOFC applications.

On-going efforts within the research groups involved here have concentrated on the synthesis of homogeneous perovskite powders via oxalate-chemistry co-precipitation of precursor powders of barium, cerium, zirconium and cobalt followed by high-temperature calcinations. In this fabrication route, co-precipitation produced finely mixed powders of the precursors which were subsequently calcined at 1550°C to achieve a homogeneous perovskite-related oxide with chemical composition close to $BaCe_{0.25}Zr_{0.60}Co_{0.15}O_{3-x}$. This material was demonstrated to have both catalytic and mixed-ionic conducting properties, thus verifying the viability of cobalt-doped barium cerate/zirconate blends for hydrogen extraction; additionally this material was observed to have substantial catalytic activity and selectivity towards hydrogen production via oxidative reforming of methanol [21].

A related report in this Proceedings details research using another synthesis route for preparing perovskite nanoparticles of barium cerates with and without cobalt doping and subsequently forming either homogeneous or heterogeneous $BaCe_{1-\alpha-\beta-\gamma}Zr_\beta Co_\gamma O_{3-x}$ materials for combined hydrogen catalytic production and electrochemical separation [22].

EXPERIMENTAL DETAILS

$BaCe_{0.25}Zr_{0.60}Co_{0.15}O_{3-x}$ (BCZC) was synthesized using a co-precipitation approach. Stoichiometric amounts of $Ba(NO_3)_2$, $Ce(NO_3)_3.6H_2O$, $Co(NO_3)_2.6H_2O$ and $ZrO(NO_3)_2.xH_2O$ were dissolved in de-ionized water. The solution was gradually added to an aqueous oxalic acid solution. Reaction mixture was aged under stirring for 2.5 h at ~ 50°C and pH ~ 2. Aqueous ammonium hydroxide solution was used to control the pH. Precipitate was washed with 1500 mL de-ionized water and 500 mL anhydrous ethanol and filtered out using a vacuum filtration set-up. Precipitate was then air-dried at ~ 140°C for 6 h. The dried precipitate was calcined on a platinum foil in an alumina crucible for 4 h at 1150°C and for 4 h at 1550°C. Materials analysis was performed to verify the phase homogeneity and to confirm the composition. X-ray powder diffraction (XRD) scans were obtained on both samples; electron diffraction patterns, X-ray energy-dispersive spectroscopy (XEDS) scans and elemental maps were obtained only on the sample calcined at 1550°C. The conductivity of the final materials has been investigated by AC impedance spectroscopy. The catalytic activity of the material for hydrogen production via dry oxidative reforming of methanol has been investigated.

RESULTS AND DISCUSSION

Observed XRD pattern for sample calcined at 1150°C (Figure 1) indicated that the material consisted of 2 phases – one based on barium cerate, the other based on barium zirconate.

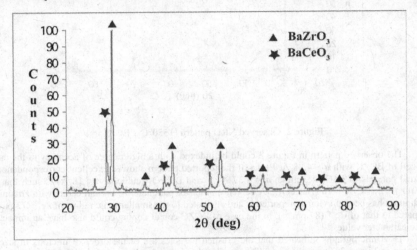

Figure 1: Powder XRD pattern from material calcined at 1150°C

For comparison, the XRD pattern shown in Figure 2 is from a sample that has been heat treated at 1550°C and corresponds well to a reported pattern of $Ba_2CeZrO_{5.5}$ (cubic, $a = 8.57$ Å) [23] Clearly, heat treatment at a higher temperature is required in order to obtain a homogeneous barium cerium zirconate phase. This observation is consistent with a previous reports [24] where it has been reported that heat treatment temperature >1500°C are required to obtain homogeneous BaCeZr phase using oxalate precipitation, spray drying and subsequent solid state reaction. The sample calcined at 1550°C will be referred to as BCZC.

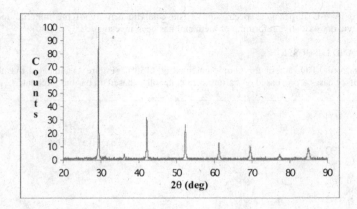

Figure 2: Observed XRD pattern (1550°C – BCZC)

The observed pattern in Figure 2 could be indexed with a high degree of accuracy to the cubic unit cell of BCZC with $a \sim 4.28$ Å. Peaks of the reported pattern showed excellent correspondence to reported patterns of $BaCeO_3$ (cubic, $a = 4.37$ Å) [25] and $BaZrO_3$ (cubic, $a = 4.18$ Å) [26] such that, a ($BaZrO_3$) < a (BCZC) < a ($BaCeO_3$). A decrease in lattice parameter of barium cerate with zirconium substitution has been previously reported [18] and attributed to the smaller ionic radius of Zr^{4+} (72 pm [27]) compared to that of Ce^{4+} (87 pm [27]). In the case of BCZC cobalt doping could also have an impact on lattice parameter values.

TEM micrographs indicated that the crushed particles are of the order of microns in size as illustrated in Figure 3. For preparing TEM samples the sintered pallet was crushed and suspended in methanol. Clearly the micrograph does not indicate the actual crystallite size. The reaction product from the Oxalate precipitation route was amorphous and there was no specific crystal size. Electron diffraction results on BCZC showed excellent correspondence to reported pattern of $BaCeO_3$. XEDS showed all 4 metal ions – Ba, Ce, Zr and Co to be present in the sample. Elemental maps of a typical BCZC particle showed the metal ions to be uniformly distributed across individual grains (Figure 4). XRD and TEM analyses indicated that BCZC was homogeneous with a cubic perovskite crystal structure.

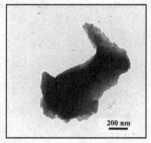

Figure 3: TEM bright-field image of a BCZC grain

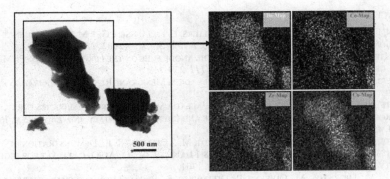

Figure 4: Elemental maps of a BCZC granular region

A heat-treatment temperature of 1550°C involves expenditure of large amount of electrical power. In addition to lowering the energy cost, decreasing the heat treatment temperature to ~1000°C would also preclude the use of Pt thereby removing a possible source of contamination. Zhong [28] has reported a synthesis method using ammonium carbonate and ammonium hydroxide as precipitant whereby heat treatment temperature required for homogeneous BaCeZr oxide phase was only 1000°C. A novel synthesis method to further reduce preparation temperature of BCZC [22] is reported elsewhere in this proceedings.

Reforming capabilities of this powder under different environments has been studied in the temperature range of 400-600 °C. Gas chromatography analysis under dry oxidative reforming environment indicated hydrogen, carbon monoxide, carbon dioxide and methane to be present in the dry reaction product mixture. Comparison of overall conductivity in different environments indicates a dominant proton conducting mechanism in this material. Details of the results of the functional aspects of the investigation have been reported elsewhere [21].

CONCLUSIONS

The authors have successfully synthesized $BaCe_{0.25}Zr_{0.60}Co_{0.15}O_{3-x}$. The precipitated material is initially amorphous and forms a single-phase crystalline materials after heat treatment at 1550 °C. Materials analysis confirm homogeneous ABO_3 or $A_2B_2O_6$ cubic phase. The functional aspect of this material has been presented elsewhere [21]. The resulting material is therefore a promising candidate for integrated hydrogen generation and electrochemical separation for high-purity hydrogen production. Future research is aimed at investigating the interactions between catalytic activity and both electronic and ionic potentials developed as a result of electrochemical hydrogen separation, and detailed permeation studies in the presence of reforming chemistries.

ACKNOWLEDGEMENTS

The authors gratefully acknowledge Peter Menard of the Connecticut Global Fuel Cell Center, Prof. William E. Mustain of the University of Connecticut and Prof. Jodie Lutkenhaus of Yale University for their assistance with electrochemical analysis. They also thank Dr Roger Ristau and Dr. Virgil Solomon for technical assistance with the EMs. AS and BAW gratefully acknowledge the support of a DuPont Young Professor Grant.

REFERENCES

1. BRUIJN, F. A. D.; PAPAGEORGOPOULOS, D. C.; SITTERS, E. F.; JANSSEN, G. J. M., THE INFLUENCE OF CARBON DIOXIDE ON PEM FUEL CELL ANODES. *J. POWER SOURCES* **2002**, *110*, 117.
2. QI, Z.; HE, C.; KAUFMAN, A., EFFECT OF CO IN THE ANODE FUEL ON THE PERFORMANCE OF PEM FUEL CELL CATHODE. *J. POWER SOURCES* **2002**, *111*, 239.
3. PAGLIERI, S. N.; WAY, J. D., INNOVATIONS IN PALLADIUM MEMBRANE RESEARCH. *SEPARATION AND PURIFICATION METHODS* **2002**, *31*, 1-169.
4. WILHITE, B. A.; SCHMIDT, M. A.; JENSEN, K. F., PALLADIUM-BASED MICRO-MEMBRANES FOR HYDROGEN SEPARATION: DEVICE PERFORMANCE AND CHEMICAL STABILITY. *IND. ENG. CHEM. RES.* **2004**, *43*, 7083.
5. WILHITE, B. A.; WEISS, S. E.; YING, J. Y.; SCHMIDT, M. A.; JENSEN, K. F., DEMONSTRATION OF 23WT% AG-PD MICROMEMBRANE EMPLOYING 8:1 $LaNi_{0.95}Co_{0.05}O_3/Al_2O_3$ CATALYST FOR HIGH-PURITY H_2 GENERATION. *ADV. MATER.* **2006**, *18*, 1701.
6. HOLLEIN, V.; THORNTON, M.; QUICKER, P.; DITTMEYER, R., PREPARATION AND CHARACTERIZATION OF PALLADIUM COMPOSITE MEMBRANES FOR HYDROGEN REMOVAL IN HYDROCARBON DEHYDROGENATION MEMBRANE REACTORS. *CATAL. TODAY* **2001**, *67*, 33.
7. WILHITE, B. A., HEAT MANAGEMENT. IN *MICROFABRICATED POWER GENERATION DEVICES*, MITSOS, A.; BARTON, P. I., EDS. WILEY-VCH: NEW YORK, 2009.
8. FOGLER, H. S., *ELEMENTS OF CHEMICAL REACTION ENGINEERING.* 3RD ED.; PRENTICE HALL: NEW JERSEY, 1999; P 182.
9. SUDA, A.; YAMAMURA, K.; MORIKAWA, A.; NAGAI, Y.; SOBUKAWA, H.; UKYO, Y.; SHINJO, H., ATMOSPHERIC PRESSURE SOLVOTHERMAL SYNTHESIS OF CERIA–ZIRCONIA SOLID SOLUTIONS AND THEIR LARGE OXYGEN STORAGE CAPACITY. *J MATER SCI* **2008**, *43*, 2258–2262.
10. CROCKER, M.; GRAHAM, U. M.; GONZALEZ, R.; JACOBS, G.; MORRIS, E.; RUBEL, A. M.; ANDREWS, R., PREPARATION AND CHARACTERIZATION OF CERIUM OXIDE TEMPLATED FROM ACTIVATED CARBON. *J MATER SCI* **2007**, *42*, 3454–3464.
11. IWAHARA, H.; UCHIDA, H.; ONO, K.; OGAKI, K., PROTON CONDUCTION IN SINTERED OXIDES BASED ON $BaCeO_3$. *J. ELECTROCHEM. SOC.* **1988**, *135*, 529.
12. IWAHARA, H.; YAJIMA, T.; HIBINO, T.; USHIDA, H., PERFORMANCE OF SOLID OXIDE FUEL CELL USING PROTON AND OXIDEION MIXED CONDUCTORS BASED ON $BaCe_{1-x}Sm_xO_{3-A}$. *J. ELECTROCHEM. SOC.* **1993**, *140*, 1687.
13. CHEEKATAMARLA, P. K.; LANE, A. M., CATALYTIC AUTOTHERMAL REFORMING OF DIESEL FUEL FOR H_2 GENERATION IN FUEL CELLS I. ACTIVITY TESTS AND SULFUR POISONING. *J. POWER SOURCES* **2005**, *152*, 256.
14. KUGAI, J.; VELU, S.; SONG, C., LOW-TEMPERATURE REFORMING OF ETHANOL OVER CeO_2-SUPPORTED NI-RH BIMETALLIC CATALYSTS FOR H_2 PRODUCTION. *CATAL. LETT.* **2005**, *3-4*, 355.
15. NISHIGUCHI, T.; MATSUMOTO, T.; KANAI, H.; UTANI, K.; MATSUMURA, Y.; SHEN, W.-J.; IMAMURA, S., CATALYTIC STEAM REFORMING OF ETHANOL TO PRODUCE HYDROGEN AND ACETONE. *APPL. CATAL. A* **2005**, *279* ((1-2)), 273-277.
16. ERRI, P.; DINKA, P.; VARMA, A., NOVEL PEROVSKITE-BASED CATALYSTS FOR AUTOTHERMAL JP-8 FUEL REFORMING. *CHEM. ENG. SCI.* **2006**, *61*, 5328.
17. KATAHIRA, K.; KOHCHI, Y.; SHIMURA, T.; IWAHARA, H., PROTON CONDUCTION IN ZR-SUBSTITUTED $BaCeO_3$. *SOL. ST. IONICS* **2000**, *138*, 91.
18. RYU, K. H.; HAILE, S. M., CHEMICAL STABILITY AND PROTON CONDUCTIVITY OF DOPED $BaCeO_3$–$BaZrO_3$ SOLID SOLUTIONS. *SOL. ST. IONICS* **1999**, *125*, 355.
19. BATISTA, M. S.; SANTOS, K. S.; ASSAF, E. M.; ASSAF, J. M.; TICIANELLI, E. A., HIGH EFFICIENCY STEAM REFORMING OF ETHANOL BY COBALT-BASED CATALYSTS. *J. POWER SOURCES* **2004**, *134*, 27.

20. SLADE, R. C. T.; SINGH, N., SYSTEMATIC EXAMINATION OF H_2 ION CONDUCTION IN RARE-EARTH DOPED BARIUM CERATE CERAMICS. *SOL. ST. IONICS* **1991,** *46*, 111.

21. SURESH, A.; BASU, J.; WILHITE, B. A.; CARTER, C. B., SYNTHESIS AND CATALYTIC ACTIVITY OF CO-DOPED $BACE_xZR_{1-x}O_3$ MIXED CONDUCTORS FOR HYDROGEN REFORMING. *MAT. RES. SOC. SYMP. PROC.* **2009.**

22. BASU, J.; WINTERSTIEN, J. P.; BHOWMICK, S.; CARTER, C. B., SYNTHESIS AND CHARACTERIZATION OF PURE, DOPED-CEO_2 AND ITS DERIVATIVES FOR ENERGY & ENVIRONMENT. *THESE PROCEDINGS* **2009.**

23. 48-0335, J. C. (ICDD POWDER DIFFRACTION FILE)

24. WIENSTRÖER, S.; WIEMHÖFER, H.-D., INVESTIGATION OF THE INFLUENCE OF ZIRCONIUM SUBSTITUTION ON THE PROPERTIES OF NEODYMIUM-DOPED BARIUM CERATES. *SOL. ST. IONICS* **1997,** *101-103*, 113.

25. 75-0431], J. C. (ICDD POWDER DIFFRACTION FILE)

26. 74-1299], J. C. (ICDD POWDER DIFFRACTION FILE)

27. CARTER, C. B.; NORTON, M. G., *CERAMIC MATERIALS: SCIENCE AND ENGINEERING*. SPRINGER: NEW YORK, 2007.

28. ZHONG, Z., STABILITY AND CONDUCTIVITY STUDY OF THE $BACE_{0.9-x}ZR_xY_{0.1}O_{2.95}$ SYSTEMS. *SOL. ST. IONICS* **2007,** *178*, 213.

IN SITU X-RAY DIFFRACTION AND RAMAN SPECTROSCOPY OF LiF-ADDED Ba(Zr$_{0.7}$Ce$_{0.1}$Y$_{0.2}$)O$_{2.9}$ CERAMICS

C.-S. Tu[a]*, S. C. Lee[a], C.-C. Huang[a], R. R. Chien[b], V. H. Schmidt[b], and C.-L. Tsai[b]
[a]Department of Physics, Fu Jen Catholic University, Taipei, Taiwan 242, R.O.C.
[b]Department of Physics, Montana State University, Bozeman, MT 59717, USA

ABSTRACT

In situ X-ray diffraction (XRD) and post Raman scattering have been used to study structure and thermal stability of lithium fluoride (LiF)-added (7 % weight ratio) Ba(Zr$_{0.7}$Ce$_{0.1}$Y$_{0.2}$)O$_{2.9}$ (BZCY712) proton conducting powders after calcination at various tempeatures in 1 atm CO$_2$. It was found that LiF begins to react with BZCY712 above 450°C upon heating and obviously enhances chemical decomposition to BaCO$_3$ and possible Ba$_3$Ce$_2$(CO$_3$)$_5$F$_2$ compounds as temperature increases. After cooling from 900°C in CO$_2$, decomposition of possible Ba(Zr$_{0.8}$Y$_{0.2}$)O$_{2.9}$ (BZY82) and (Zr,Ce,Y)O$_2$ structure (or high-cerium BZCY) were observed in the LiF-added BZCY712. The XRD intensity of possible Ba$_3$Ce$_2$(CO$_3$)$_5$F$_2$ grows significantly upon cooling from 900°C. The post XRD and Raman spectra of 7 wt% LiF-added BZCY712 powders calcined at 1200°C in air for 5 hrs, do not show existence of Ba$_3$Ce$_2$(CO$_3$)$_5$F$_2$ or LiF-related compounds. This work suggests that LiF and LiF-related chemical compounds can be removed after calcining or sintering at high temperature ($\geq$1200°C).

INTRODUCTION

One important issue facing proton conducting ceramics for hydrogen purification and solid oxide fuel cell (SOFC) electrolytes is the chemical instability to the environment, especially reaction with coal syngas components such as CO$_2$, H$_2$S, and other trace species.[1,2] However, the effect of coal syngas species on SOFC related materials is not presently well known. Though proton conductors are promising candidates for SOFCs because of their low activation energy,[3-6] the major challenge for these materials is to find a proper compromise between ionic conductivity and chemical stability in various environments and operation temperature ranges. BaCeO$_3$ has been known to exhibit high ionic conductivity at high temperatures ($\geq$ 500°C) but poor chemical stability was observed in CO$_2$ and H$_2$O atmospheres.[3-9] On the other hand, yttrium-doped BaZrO$_3$ shows both sufficient chemical and thermal stability.[10,11] Zr substitution for Ce can improve chemical stability but decreases the ionic conductivity. Partially substituting Zr for Ce reduces its tendency to decompose into BaCO$_3$ and other oxides in a CO$_2$-containing environment at intermediate temperatures (typically 700-850°C).

Recent post X-ray diffraction (XRD) results of Ba(Ce$_{0.8}$Y$_{0.2}$)O$_3$ (BCY82) and Ba(Zr$_{0.1}$Ce$_{0.7}$Y$_{0.2}$)O$_{3-\delta}$ (BZCY172) powders, suggested that BCY20 can easily decompose to BaCO$_3$, CeO$_2$ and Y$_2$O$_3$ after exposure to 2% CO$_2$ (with H$_2$) at 500°C.[12] BZCY172 exhibited good chemical and kinetic stability in 2% CO$_2$-containing atmosphere below 500°C.[12] Our recent *in situ* XRD and post Raman results show that BZCY712 exhibits very promising thermal stability in 1 atm CO$_2$ up to 900°C (at least) without obvious chemical decomposition.[13]

Lithium fluoride (LiF) has been used as an aid to reduce sintering temperature and also to enhance mass density in proton conducting ceramics. However, there is no much knowledge of the effect regarding the LiF addition in proton conducting ceramics. In this work, *in situ* temperature-dependent XRD was employed to investigate the LiF-added effect (7 wt%) on thermal stability (or chemical decomposition) of calcined Ba(Zr$_{0.7}$Ce$_{0.1}$Y$_{0.2}$)O$_{2.9}$ powders in 1 atm CO$_2$ in the region of 20-900°C. The post XRD and Raman scattering spectra of 7 wt% LiF-added BZCY712 powders were also carried out after calcining at 1200°C for 5 hrs, and after exposure to 1 atm CO$_2$ at 850°C (after a previous calcination at 1200°C in air).

X-Ray Diffraction and Raman Spectroscopy of LiF-Added Ba(Zr$_{0.7}$Ce$_{0.1}$Y$_{0.2}$)O$_{2.9}$ Ceramics

EXPERIMENTAL

Ba(Zr$_{0.7}$Ce$_{0.1}$Y$_{0.2}$)O$_{2.9}$ (BZCY712) powders were synthesized by the glycine-nitrate process and then were calcined at 1300°C in air for 5 hours to ensure a single perovskite phase. The LiF powders (7% weight ratio or 7 wt%) was added in calcined BZCY712 powders during the milling process. For *in situ* X-ray diffraction measurements, a high-temperature Rigaku Model MultiFlex X-ray diffractometer with Cu Kα_1 (λ = 0.15406 nm) and Cu Kα_2 (λ = 0.15444 nm) radiations was used. LiF-added BZCY712 powders were placed and smoothly pressed on the platinum sample holder. The powder samples were scanned at room temperature first before flowing CO$_2$, and then the atmosphere was switched to flowing 1 atm pure CO$_2$. The temperature was raised in steps from room temperature. Each XRD scan was taken after holding the sample at the setting temperature for about 10 minutes to ensure a complete reaction with CO$_2$. The same scan process was repeated up to 900°C and then the sample was cooled to room temperature in steps. The post XRD and Raman scattering spectra of 7 wt% LiF-added BZCY712 powders were also measured after calcining at 1200°C for 5 hrs, and after exposure to CO$_2$ at 850°C (after a previous calcination at 1200°C in air).

A double grating Jobin Yvon Model U-1000 double monochromator with 1800 grooves/mm gratings and a nitrogen-cooled CCD as a detector were employed for the post micro-Raman scattering. A Coherent Model Innova 90 argon laser with wavelength λ = 514.5 nm was used as an excitation source. The Raman scattering was performed in the region of 150-1600 cm^{-1}. The LiF-added BZCY712 powders used in the post Raman scattering measurements were previously exposed to 1 atm CO$_2$ during *in situ* temperature-dependent XRD.

RESULTS AND DISCUSSION

Figure 1(a) shows the temperature-dependent XRD spectra of as prepared LiF-added BZCY712 powders in 1 atm CO$_2$-flowing atmosphere. To identify possible chemical decomposition, XRD spectrum of BaCO$_3$ (99.9% purity) was obtained at room temperature as shown in Fig. 1(b). The XRD spectrum of cebaite-(Ce) Ba$_3$Ce$_2$(CO$_3$)$_5$F$_2$ is given in Fig. 1(b). At 20°C, the main diffraction peaks of LiF-added BZCY712 powders include (110), (200), (211), (220), and (310), suggesting a body-centered cubic (bcc) unit cell according to the structure-factor calculation,[14] but the weak (111) peak is consistent with the expected simple-cubic (sc) perovskite structure. The cubic lattice parameter was estimated from the (110) peak (2θ = 29.62°) and is a = 4.262 Å.

Upon heating in 1 atm CO$_2$ atmosphere, the XRD spectra remain the same below 450°C without obvious chemical decomposition. However, near 450°C [Fig. 1(a)] an obvious peak appears near 2θ = 23.5° as denoted by "+", indicating appearance of BaCO$_3$ as compared with the XRD spectrum of BaCO$_3$ (99.9% purity) in Fig. 1(b), whose strongest 2θ peak appears near 24.0°. In addition, a weak peak was observed at about 28.5° as indicated by "*", possibly corresponding to Ba$_3$Ce$_2$(CO$_3$)$_5$F$_2$ according to the XRD spectrum of Ba$_3$Ce$_2$(CO$_3$)$_5$F$_2$ [Fig. 1(b)]. As temperature increases, the relative intensities of "+" and "*" peaks grow gradually up to 775°C and then essentially vanish near 800°C. In addition, a low-2θ shoulder peak clearly develops at the left-hand side of the main (110) peak upon heating and cooling. It likely corresponds to the (Zr,Ce,Y)O$_2$ structure or high-cerium BZCY, because cerium has a larger radius. Upon cooling, as seen in Fig. 1(a) the 27.4° peak of the possible Ba$_3$Ce$_2$(CO$_3$)$_5$F$_2$ structure was enhanced significantly and its intensity is even higher than the main (110) peak at 500°C. However, the almost equally strong 22.2° peak of this structure, and other peaks of this structure, do not appear at all. Upon cooling, the BaCO$_3$ structure reappears near 650°C, however its intensity is much weaker than the possible Ba$_3$Ce$_2$(CO$_3$)$_5$F$_2$ peak at 27.4°.

Figures 2(a) and 2(b) show the post Raman spectra of pure BZCY712 (without LiF) and 7 wt% LiF-added BZCY712 powders respectively after temperature-dependent exposure to 1 atm CO$_2$. To identify BaCO$_3$ through atomic vibration modes, Raman spectrum of BaCO$_3$ powders was measured at room temperature as shown in Figure 2(c). The main vibrations of BaCO$_3$ include 225, 690, 1059, and 1419 cm^{-1}. Ba$_3$Ce$_2$(CO$_3$)$_5$F$_2$ has two major Raman vibrations of 265 and 1085 cm^{-1} as given in Fig. 2(d). As indicated by "+" in Fig. 2(a), Raman vibrations 690, 1059, and 1419 cm^{-1} of BaCO$_3$ appear in the pure BZCY712 powders after *in situ* exposure to CO$_2$. In addition, 270 and 1090 cm^{-1} vibrations of the Ba$_3$Ce$_2$(CO$_3$)$_5$F$_2$ structure [as indicated by "*" in Fig. 2(b)] were also detected in 7 wt% LiF-added BZCY712 powders after an *in situ* exposure to CO$_2$. This Raman result lends stronger support to presence of the cebaite structure than is provided by the XRD spectrum alone.

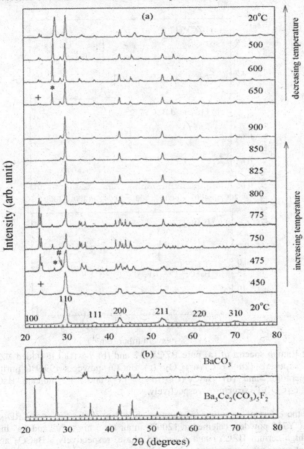

Figure 1 *In situ* X-ray diffraction spectra of (a) 7 wt% LiF-added BZCY712 upon heating and cooling in 1 atm CO$_2$, and (b) X-ray diffraction spectra of BaCO$_3$ and Ba$_3$Ce$_2$(CO$_3$)$_5$F$_2$ powders taken at room temperature. "+", "*", and "#" indicate BaCO$_3$, Ba$_3$Ce$_2$(CO$_3$)$_5$F$_2$, and (Zr,Ce,Y)O$_2$ structure (or high-cerium BZCY), respectively.

To compare thermal stability in CO_2, Fig. 3 shows the post X-ray diffraction spectra of pure BZCY712 and 7 wt% LiF-added BZCY712 after an *in situ* exposure (20-900°C). It reveals that LiF reacted with BZCY712 powders upon heating and that the possible Ba$_3$Ce$_2$(CO$_3$)$_5$F$_2$ structure was significantly enhanced after an *in situ* exposure (20-900°C) to CO_2. The 7 wt% addition of LiF in BZCY712 also induces decomposition of possible BZY82 and (Zr,Ce,Y)O$_2$ structure (or high-cerium BZCY) as evidenced by the low-2θ shoulder beside the main (110) peak near 2θ = 30°.

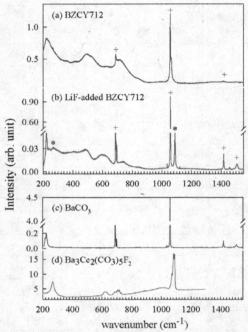

Figure 2 Post Raman spectra of (a) pure BZCY712 and (b) 7 wt% LiF-added BZCY712 after an *in situ* exposure (20-900°C) to CO_2, (c) BaCO$_3$ powders (99.9% purity) taken at room temperature, and (d) Ba$_3$Ce$_2$(CO$_3$)$_5$F$_2$. "+" and "*" indicate BaCO$_3$ and possible Ba$_3$Ce$_2$(CO$_3$)$_5$F$_2$ structures, respectively.

To study the effect of calcining temperature, Fig. 4(a) shows the post XRD spectrum of 7 wt% LiF-added BZCY712 powders calcined at 1200°C in air for 5 hrs. "#" and "□" indicate (Zr,Ce,Y)O$_2$ structure (or high-cerium BZCY) and a second phase, respectively. BaCO$_3$ and Ba$_3$Ce$_2$(CO$_3$)$_5$F$_2$ structure do not occur after exposure to air at 1200°C and were not detected either in the post Raman spectrum as shown in Fig. 5(a). Fig. 4(b) is the post XRD spectrum after exposure to CO_2 at 850°C for 7 wt% LiF-added BZCY712 powders, which were previously calcined at 1200°C in air for 5hrs. As expected in CO_2 atmosphere, BaCO$_3$ was observed as indicated by "+" in Fig. 4(b) and was also confirmed by the Raman vibrations of 690, 1059, and 1419 cm^{-1} as shown in Fig. 5(b). In addition,

(Zr,Ce,Y)O$_2$ structure (or high-cerium BZCY) was detected in Figs. 4(a) and (b). Ba$_3$Ce$_2$(CO$_3$)$_5$F$_2$ structure was not observed after calcining at 1200°C in air or exposure to CO$_2$ at 850°C (after calcining at 1200°C) as evidenced in Figs. 4 and 5. This implies that LiF and LiF-associated chemical compounds can be removed after a calcining (or sintering) process at high temperature (≥ 1200°C).

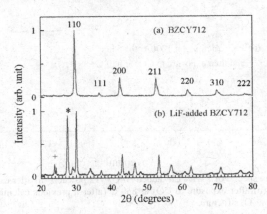

Figure 3 Post X-ray diffraction spectra of (a) pure BZCY712 and (b) 7 wt% LiF-added BZCY712 after an *in situ* exposure to CO$_2$ (20-900°C). "+" and "*" indicate BaCO$_3$ and Ba$_3$Ce$_2$(CO$_3$)$_5$F$_2$ structures, respectively.

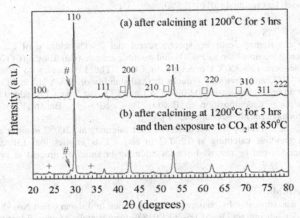

Figure 4 Post XRD spectra of 7 wt% LiF-added BZCY712 powders (a) after calcining at 1200°C for 5 hrs and (b) after exposure to CO$_2$ at 850°C (after a previous calcining at 1200°C in air). "+" and "#" indicate BaCO$_3$ and (Zr,Ce,Y)O$_2$ structure (or high-cerium BZCY), respectively. "□" represents a second phase.

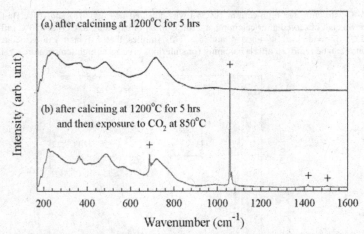

Figure 5   Post Raman spectra of 7 wt% LiF-added BZCY712 powders (a) after calcining at 1200°C for 5 hrs and (b) after exposure to CO_2 at 850°C (after a previous calcining at 1200°C).   "+" indicates $BaCO_3$ structure.

Nuclear Reaction Analysis (NRA) (which is sensitive to Li content) also shows negligible residue of LiF-related chemicals in $Ba(Zr_{0.6}Ce_{0.2}Y_{0.2})O_{2.9}$ (BZCY622) for samples sintered for various dwell times at 1400°C, with temperature ramp rates +5°C/min and -10°C/min.   Longer dwell times at 1400°C yield much less residual LiF (≤ 0.5%).

CONCLUSIONS
The XRD and Raman scattering spectra reveal that 7 wt% addition of LiF in BZCY712 can enhance chemical decomposition to $BaCO_3$ and possible cebaite-(Ce) $Ba_3Ce_2(CO_3)_5F_2$ after an *in situ* temperature-dependent (20-900°C) exposure to CO_2.   The LiF begins to react with BZCY712 powders above 450°C and obviously reduces thermal stability of BZCY712 in CO_2 atmosphere upon heating.   After an *in situ* exposure (20-900°C) to 1 atm CO_2, the 7 wt% LiF-added BZCY712 exhibits an obvious chemical decomposition to $BaCO_3$, $Ba_3Ce_2(CO_3)_5F_2$, $Ba(Zr_{0.8}Y_{0.2})O_{2.9}$ (BZY82), and $(Zr,Ce,Y)O_2$ structures (or high-cerium BZCY).

$Ba_3Ce_2(CO_3)_5F_2$ structure was not observed after calcining at 1200°C in air or exposure to CO_2 at 850°C (after a previous calcining at 1200°C in air).   This implies that LiF and LiF-associated chemical compounds can be removed after a calcining or sintering process at high temperature (≥ 1200°C).

ACKNOWLEDGEMENT
This work was supported by National Science Council of Taiwan Grant No. 96-2112-M-030-001, and by DOE under subcontract DE-AC06-76RL01839 from Battelle Memorial Institute and PNNL.

REFERENCES
1. J.P. Trembly, R.S. Gemmen, and D.J. Bayless, J. Power Sources 163 (2007) 986-996.
2. R.S. Gemmen and J. Trembly, J. Power Sources 161 (2006) 1084-1095.
3. S. McIntosh and R.J. Gorte, Chem. Rev. 104 (2004) 4845-4865.
4. S.M. Haile, Acta Materialia 51 (2003) 5981-6000.
5. C.W. Tanner and A.V. Virkar, J. Electrochem. Soc. 143 (1996) 1386-1389.
6. S.V. Bhide and A.V. Virkar, J. Electrochem. Soc. 146 (1999) 2038-2044.
7. K. H. Ryu and S. M. Haile, Solid State Ionics 125 (1999) 355-367.
8. G. Ma, T. Shimura, and H. Iwahara, Solid State Ionics 110 (1998) 103-110.
9. K. Katahira, Y. Kohchi, T. Shimura, and H. Iwahara, Solid State Ionics 138 (2000) 91-98.
10. F.M.M. Snijkers, A. Buekenhoudt, J. Cooymans, and J.J. Luyten, Scripta Materialia 50 (2004) 655-659.
11. A. Magrez and T. Schober, Solid State Ionics 175 (2004) 585-588.
12. C. Zuo, S. Zha, M. Liu, M. Hatano, and M. Uchiyama, Advanced Materials 18 (2006) 3318-3320.
13. C.-S. Tu, R. R. Chien, V. H. Schmidt, S.-C. Lee, C.-C. Huang, and C.-L. Tsai, submitted to Journal of Applied Physics (2008).
14. B.D. Cullity, Elements of X-ray diffraction, (Addison-Wesley Publishing, 1978).

Seals

SEALING GLASSES FOR SOFC – DEGRADATION BEHAVIOUR

Jochen Schilm, Axel Rost, Mihails Kusnezoff, Alexander Michaelis

Fraunhofer Institute for Ceramic technologies and systems
Winterbergstr. 28, 01277 Dresden, Germany

ABSTRACT

The stability of sealing glasses is one of the critical issues concerning the long-term operation of SOFC. Different reasons are responsible for failures of the glass sealings resulting from thermal cycling, stack deformation or chemical degradation due to interactions with fuel gas, interconnectors or protective coatings. Different reactions with an interconnector material (Crofer 22) under oxidizing and reducing conditions lead to crystalline product layers, which affect the bonding strength and microstructure of the sealing material at the interfaces. Effects of externally applied electric voltage on the sealing glass provide insights into mechanisms of reactions at anodically and cathodically polarized interfaces of the interconnector material. Tests of model sealings under electrical voltage in combination with a dual atmosphere reveal additional information about the stability of sealing glasses.

INTRODUCTION

Studies on the development of sealing glasses for SOFC often consider intrinsic properties of the glasses such as the TEC, crystalline phases, viscosity and short time chemical compatibility with metallic and ceramic substrates [1,2,3,4]. In several studies also the long term stability of the intrinsic glass properties is regarded as well as the behaviour in contact with substrates at high temperatures [4,5,6,7]. However the results of such tests do not allow reliable statements about the long term stability and the degradation behaviour of sealing glasses under operating conditions of SOFCs. Only a few studies regard the facts that SOFCs sealing glasses are exposed to a dual atmosphere and, depending on the design, to the electrical voltage [8,9,10]. No studies are published about the interfacial reactions of SOFC sealing glasses in contact with metallic interconnector materials under the influence of electrical voltage in air or in dual atmosphere. A reliable method to investigate these complex reactions between glass, interconnector materials and the atmosphere is given by tests of model sealings under conditions comparable to the operating conditions of SOFC. Model sealings can be prepared from glass powders which are processed to pastes or tapes. For the investigation of basic interfacial reactions between glass and metal without interfering effects of pores or bubbles in the glass it is convenient to use sandwich arrangements of bulk glass samples between two conducting electrodes. As electrode materials either reactive interconnector materials or inert gold foils can be employed. These samples are exposed to high temperatures in an oxidizing atmosphere. In both cases an electric potential can be applied across the sealing glass at high temperatures to simulate the electrical conditions in a SOFC.

METHODS & EXPERIMENTALS

All experiments have been performed with a partial crystallizing SOFC sealing glass based on the system SiO_2-Al_2O_3-BaO with additions of B_2O_3 and ZnO. The only crystalline phase in this glass is sanbornite (BS2). The main application of this sealing glass is SOFC with operating temperatures of 700-850°C. The glasses were molten from reagent grade raw materials (SiO_2, $BaCO_3$, $Al(OH)_3$, H_3BO_3 and ZnO) in a Pt-crucible at 1500°C/1h. For the preparation of powders the glasses were fritted into water after melting, dried at 150°C for 2h and finally grinded in a jet mill.

For investigations of the degradation behaviour in dual atmosphere tapes with a green thickness of 500µm were cast from the glass powder. Model sealings were prepared from these glass tapes in a sandwich arrangement of two coupons of Crofer 22 where a ring of the glass tape with a defined geometry was placed in between (fig. 1a). The two adjacent holes in one coupon allow a continuous gas flow inside the model sealing (fig. 1c). The furnace allows simultaneous tests of 4 model sealings in a dual atmosphere while an electrical voltage is applied across the sealing. All prepared samples were tested for their Helium-leakage rate and found to be dense with leakage rates lower than 10^{-8} l mbar s^{-1} cm^{-1}. During the experiments the electrical resistance of the samples are registered continuously with an experimental setup according to figure 1c. A gas mixture consisting of 81% N_2, 9% H_2, 7% CO_2 and 3% H_2O was used on the fuel side of the dual atmosphere. Within this work a term "cathode" is used for electrochemically reducing conditions while the term "anode" stands for

185

oxidizing conditions. Additionally melts of the glass were cast into bars and tempered below Tg to remove internal tensions. From these bars compact glass samples in the dimensions of $5,0 \times 5,0 \times 1,0 \text{ mm}^3$ were prepared (Fig. 1b). There main surfaces were grinded down to a main roughness below 2 µm. These glass samples were used in sandwich arrangements between Crofer 22 or gold plates as electrode materials according to Figure 1d to study interfacial reactions and formation of pores without superimposing effects of existing pores in glass sealings which are derived from glass powders. The lateral dimensions of the compact glass samples correspond nearly with the broadness of the glass sealings prepared from glass tapes. These samples were tested at 850°C in a tube furnace in air with the same electrical setup as the model sealings (Fig. 1d).

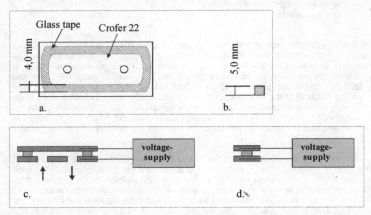

Fig. 1   a. Top view of módel sealings prepared from glass tapes; b. Test sample prepared from compact glass bars; c. experimental setup for electrical measurements of model sealings; d. experimental setup for electrical measurements of compact glasses

RESULTS & DISCUSSION

a. Interactions of compact glass samples with Crofer 22 in air

Sandwich arrangements of compact glass slides in contact with reactive electrode material (Crofer 22) or inert electrode material (gold) were annealed in air for 100 h under an electrical load of 0,7V (DC). Gold as an electrode material was found to undergo no interfacial reactions with the sealing glass. For the purpose of separating anodic and cathodic effects of interfacial reactions between Crofer 22 and glass, Crofer 22 was replaced by inert gold electrodes (500 µm foils). Samples with four different arrangements of the electrodes have been tested. In Figure 2 the results of the electrical measurements are shown. The time dependent progress of the electrical resistances of all samples is similar and after reaching a nearly constant level the specific resistances range from $0,5 \cdot 10^6$ to $3,0 \cdot 10^6$ Ωcm. In the following the results of SEM investigations of the different interfaces are shown and discussed with respect to characteristic features of the electrical measurements.

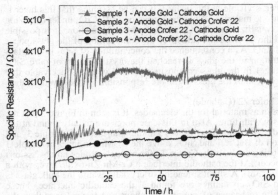

Fig. 2 Specific resistances of crystallized compact glass samples between two metallic electrodes (Gold or Crofer 22) during aging in air at 850°C with an externally applied voltage of 0,7 V

Sample1 - Gold (Anode) – Gold (Cathode):
For the combination of gold for both electrodes no additional crystalline products but BS2 were found at the anodic and cathodic interfaces (fig. 3a, 3e and 3f). The formation of pores was located at both interfaces but not inside the glass (fig. 3e and 3.f). The electrical resistance remained nearly constant after 20h of testing. Within the first 20h of measurement strong fluctuations of the electrical resistance were registered (fig. 2). It is assumed that these strong variations of the resistance are connected with the formation of pores during this period. The formation of pores was considerable stronger at the anode than at the cathode. Due to the absence of any reaction products at the interfaces the specific electrical resistance of $1,3 \cdot 10^6$ Ωcm represents the ohmic resistance of the crystallized glass phase. An effect of the pores at the interfaces on this value cannot be quantified.

Sample 2 - Gold (Anode) – Crofer 22 (Cathode):
When the cathode material is changed from Gold to Crofer 22 also within the first 20h of the experiment strong fluctuations of the resistance were measured (Fig. 2). As well in this case an extensive formation of pores is located at the anode (Fig. 3b). Considerably fewer pores are found at the cathode and no pores were present in the bulk glass. At the cathode a dense crystalline layer consisting of a Cr-Mn-Zn-Oxide of unknown stoichiometry and a thickness of about 5 µm was formed (Fig. 3g). Small pores were located beneath the oxide layer indicating an interaction of the Crofer 22. Single crystallites of the oxide layer were distributed in the bulk glass. $BaCrO_4$ was only detected at the outer edges of the glass phase on the cathodically polarized interface where an oxidation of the Crofer 22 is possible (Fig. 3b). No additional product layers are found at the anode (fig. 3h). Sample 2 showed the highest specific resistance of all samples tested within this series. But due to the strong formation of pores at the anode it must be assumed that a decreasing contact area between the glass and the anode is responsible for the high resistance values.

Sample 3 - Crofer 22 (Anode) – Gold (Cathode):
For the next experiment the anode material was changed from Gold to Crofer 22. Now within the initial period of testing a smooth increase of the electrical resistance is registered (fig. 2). A constant level at about $0,5 \cdot 10^6$ Ωcm is reached after nearly 25 hours. As seen in figure 3.c the formation of pores at the anode and the cathode is much less pronounced than in the samples 1 and 2. No pores are found inside the bulk glass. It has to be remarked that using this arrangement of electrodes, always an abscission between glass and Crofer 22 but not between glass and gold occurred. In this case at the anodic interface different reaction products of glass and Crofer 22 are detected. An about 5µm thick and dense layer of a Cr-Mn-Zn-Oxide is located on the surface of the Crofer 22 (Fig. 3k). Above this layer a mixture of $BaCrO_4$ and SiO_2 has been formed throughout the whole interface and not only the outer edges of the glass phase as seen in sample 2 (Fig. 3k). The detached anode of these samples is seen in

figure 3l, where the Cr-Mn-Zn-oxide layer is visible at the surface. It can be assumed that this layer is responsible for the abscission of the electrode material. While the formation of $BaCrO_4$ under oxidizing conditions at the anode is evident, the crystallization of SiO_2 needs further explanation. It is possible that due to the formation of $BaCrO_4$ and the Cr-Mn-Zn-oxide layer the composition of the glass phase in this region becomes poor on BaO and ZnO. As a consequence the concentration of SiO_2 in the glass increases. When a critical SiO_2-concentration in the glass is reached the crystallization of pure SiO_2 occurs what in turn leads to further changes in the glass composition. But it is also conceivable that additional effects, coming from the polarization of the glass phase are responsible for the crystallisation of SiO_2 at the anode.

Sample 4 - Crofer 22 (Anode) – Crofer 22 (Cathode):

For the last sample only Crofer 22 is chosen as material for the electrodes. It is seen in Figure 2 that the electrical resistance increases continuously with time. Also in this case pores are mainly found at the anodic interface of the sample (Fig. 3d). A comparison of the anodic interfaces of sample 3 and sample 4 reveal a much stronger formation of $BaCrO_4$ and SiO_2 above the Cr-Mn-Zn-layer in sample 3 (Fig. 3k) than in sample 4 (Fig. 3n). By using exclusively Crofer 22 for both electrodes no abscissions between the glass phase and the anode occur. At the cathodic interface a Cr-Mn-Zn-oxide layer with a maximum thickness of 2µm is formed (Fig. 3m). A slight spreading of crystallites into the bulk glass is observed. It has to be remarked that no SiO_2 crystallites are found at the cathodic interface. These results agree with the fact that the specific resistance of sample 3 is about two times lower than the specific resistance of sample 4. In case of sample 3 the formation of larger amounts of $BaCrO_4$ can cause a higher current.

Overview	interface in higher magnification

a. Gold (Anode) – Gold (Cathode)

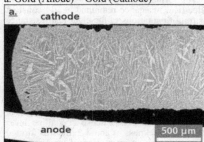

Gold (Anode) – Crofer 22 (Cathode)

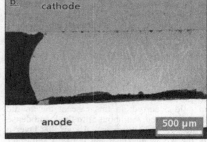

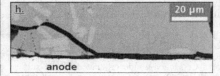

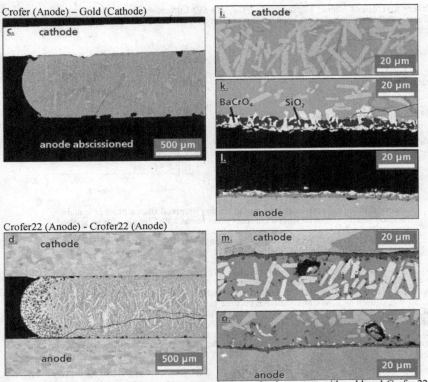

Fig. 3   SEM images of interfaces of compact glass samples in contact with gold and Crofer 22 after aging in air at 850°C with an externally applied voltage of 0,7 V

b. Interaction with fuel gas (behaviour of glass under dual atmosphere)

To study effects of dual atmospheres in combination with an applied electrical voltage on the interfacial reactions of sealing glass A with Crofer 22, model sealings were tested in the dual atmosphere furnace at 850°C. After sealing the glass contains up to 20 Vol.-% of pores but nevertheless possesses closed porosity (no leakage). The pores have to be considered as a part of the microstructure of the crystallized glass. Thus in comparison with a compact glass, a porous sealing glass can have differing properties. This is exemplary seen in Figure 4a where the specific resistances (850°C in air) of a model sealing prepared from a glass tape is compared with a compact glass sample. The specific resistance of glass tape differ strongly from the specific resistance of the compact glass sample.

Results of tests with model sealings in air and in a dual gas atmosphere in combination with an externally applied voltage of 0,7 V are shown in Figure 4b. The time dependent characteristics of the specific electrical resistances are typical and are measured reproducibly with many samples. Foremost a sharp increase of the resistance occurs. After about 100h of testing a maximum value is reached and afterwards the specific resistance decreases slightly with time. It is seen that the chosen atmospheres cause no strong differences in the measured resistances. After 300h of testing values the sealing glass shows still specific resistances between $4 \cdot 10^6$ and $6,5 \cdot 10^6$ Ω cm which satisfy the requirements for the insulating properties of sealing glasses for SOFC. Also SEM investigations of these samples reveal no significantly differing appearances between the interface layers which can be attributed to the varying fuel gas atmospheres.

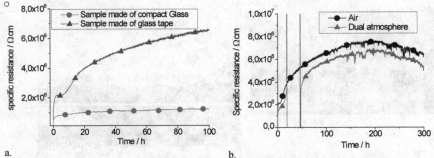

a.   b.

Fig. 4   a. Specific electrical resistance of a model sealing prepared from a glass tape and
              a glass sample prepared from a compact glass bar at 850°C in air with an applied
              voltage of 0,7 V
            b. Specific electrical resistances of model sealings at 850°C in air and in dual
              atmosphere with an applied voltage of 0,7 V

In further tests related with the stability of the glass in a dual atmosphere the applied external voltages across the glass phase have been increased from 0,7 V to 1,0 V, 1,3 V and also 30 V. Figure 5 shows the specific resistances of 4 samples, measured under these modified conditions in a dual atmosphere at 850°C. 30 V is chosen as a maximum voltage because this value corresponds to the open circuit voltage of a SOFC-stack consisting of 30 repeating units. In Figure 4 is seen that maximal resistance reached after shorter times of testing applied voltage is increased (120 h at 0,7 V and 1,0 V, 80 h at 1,3 V and 30 h at 30 V). Also the decreasing rate of the resistances after passing the maximum value increases with higher voltages, indicating different changes of the glass structure or of the interface layer at the electrodes.

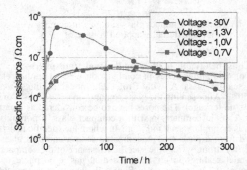

Fig. 5   Specific electrical resistances of model sealings at 850°C in a dual atmosphere with different
            voltages applied across the samples.

For this reason the microstructures of the glass and the interfaces of sample 1 (0,7 V) and sample 4 (30 V) according to figure 5 are investigated by SEM. The formation of BaCrO$_4$ occurred only at the 3-phase boundary of the glass on the air side (Fig. 6). It seems that the BaCrO$_4$ growths along the glass-metal interface into the inner region of the glass sealing. At the 3-phase boundary on the fuel-side under reducing conditions no BaCrO$_4$ is detected.

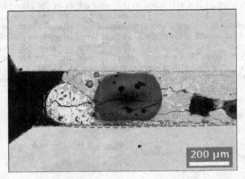

Fig. 6   SEM image of the cross section of a model sealing which shows the formation of BaCrO$_4$ (encircled) at the outer edge of the sealing glass (air side) after treatment in dual atmosphere at 850°C

Figure 7 shows the inner regions of the anodic interfaces of sample 1 (fig. 7.a) and sample 4 (Fig. 7b). The interfaces are formed by double layers consisting of Cr-Mn-Zn-oxide, which is in contact with the Crofer 22, and crystalline SiO$_2$, which is in contact with the sealing glass. It is seen that the interface layers of sample 4 (30 V) are thicker than the layers of sample 1 (0,7 V). Especially the SiO$_2$-crystallites are formed with high aspect ratios. Additionally the Cr-Mn-Zn-oxide layer of sample 4 shows pores. In contrast to the results obtained from the compact glass samples in this case the formation of crystalline SiO$_2$ at the anodically polarized interfaces occurs despite the lack of a BaCrO$_4$-formation. This excludes the assumption that a local change in the glass composition due to the crystallization of BaCrO$_4$ is exclusively responsible for the crystallisation of SiO$_2$.

 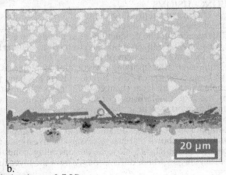

a.   b.
Fig. 7   a. Anode of sample 1 from figure 5 (applied voltage: 0,7 V)
b. Anode of sample 4 from figure 5 (applied voltage: 30 V)
Both sample were heat treated in a dual atmosphere at 850°C for 300h

The excessive formation of $BaCrO_4$ at the anodic interfaces of the compact samples can be explained by a higher surface to volume ratio of these samples which is exposed to air. In contrast the model sealing has a different design and as a consequence only one boundary is exposed to air when these samples are tested in a dual atmosphere. Hence the compact glass samples offer a shorter diffusion path for oxygen and therefore a stronger crystallisation on $BaCrO_4$ can occur. These results support the assumption, that the formation of $BaCrO_4$ starts at the outer edges of the glass, which are exposed to air and proceeds along the anodic interface into the inner regions of the sealing. The depletion of BS2 crystallites in the bulk glass in Figure 7b in comparison to Figure 7a may be a consequence of the increased voltage applied across the glass sealing. Further work is necessary to clarify this behaviour.

At the cathodic interfaces of sample 1 Fig. 8a) and sample 4 (Fig. 8b) layers of Cr-Mn-Zn-oxide are found. The layer of sample 4 (30 V) is about three to four times thicker for sample 1 (0,7 V). The layer of sample 4 offers a dense and non porous appearance. Due to the small thickness of the layer of sample 1 an appraisement of its structure is not possible. A spreading of crystallites of the Cr-Mn-Zn-oxide into the glass is found in sample 3 and sample 4. Also no depletion of BS2-crystallites in the bulk structure of the glass is seen.

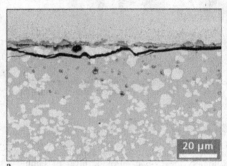 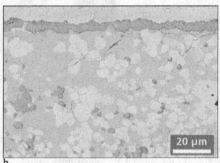

a.   b.
Fig. 8   a. Cathode of sample 1 from figure 5 (applied voltage: 0,7 V)
          b. Cathode of sample 4 from figure 5 (applied voltage: 30 V)
          Both sample were heat treated in a dual atmosphere at 850°C for 300 h

CONCLUSIONS
          Investigations of reactions of a Ba-Al-Si-glass (with additions of B_2O_3 and ZnO) with Crofer 22 under the influence of electrical voltage in air and in dual atmosphere have shown the formation of different crystalline product layers at anodically and cathodically polarized glass-metal-interfaces. Thin layers of a Cr-Mn-Zn-oxide are formed at all interfaces, what indicates a reactivity of the Crofer 22 with the ZnO component of the glass. These layers do not exceed a thickness of 10μm after 1000h of testing even after applying a voltage of 30 V across the glass. Additionally crystalline SiO_2 of similar thicknesses as the Cr-Mn-Zn-oxide layers has been formed at anodically polarized interfaces. Possible explanations are local changes in the glass composition near the interfaces or additional effects caused by the anodic polarization of the glass structure. The formation of $BaCrO_4$ has been found to occur at anodic interfaces near the 3-phase boundary of glass, air and Crofer 22. The growth of $BaCrO_4$ proceeds along the 2-phase boundary of glass and Crofer 22 into the inner region of the sealing glass. A comparison of results obtained from glass samples aged in air and from sealings treated in a dual atmosphere shows that it is necessary to investigate the mechanisms of such interfacial reactions under conditions which are close to the operating conditions of SOFC. Pure oxidizing conditions lead to product layers which differ strongly from the layer formed in the reducing atmosphere.

REFERENCES
[1] J. W. Fergus, Sealants for solid oxide fuel cells, Journal of Power Sources 147 (2005) 46–57
[2] K.L. Ley, M. Krumpelt, R.Kumar, J.H. Meiser and I. Bloom, Glass-ceramic sealants for solid oxide fuel cells: Part I. Physical properties, J. Mater. Res., 1996, 11 (6), 1489 - 1493
[3] K. Eichler, G. Solow, P. Otschik and W. Schaffrath, BAS (BaO.Al$_2$O$_3$.SiO$_2$)-glasses for High Temperature Applications, J. Eur. Cer. Soc., 1999, 19, 1101-1104
[4] S.T. Reis and R.K. Brow, Designing Sealing glasses for SOFC, J. Mat. Eng. and Perf., 15 (4), 2006, 410-413
[5] Yang, J.W. Stevenson and K.D. Meinhardt, Chemical Interactions of Barium-Calcium-Aluminosilicate-Based sealing glasses with oxidation resistant alloys, Solid state Ionics, 2003, 160, 213-225
[6] F. Smeacetto, M. Salvo, M. Ferraris, V. Casalegno, P. Asinari and A. Chrysanthou, Characterization and performance of glass–ceramic sealant to join metallic interconnects to YSZ and anode-supported-electrolyte in planar SOFCs, J. Eur. Cer. Soc, 2008, 2521 - 2527
[7] S.-B. Sohn, S.-Y. Choi, G.-H. Kim and H.-S. Song, Stable sealing glass for planar solid oxide fuel cell, J. of Non-Cryst. Solids, 2002, 297, 103-112
[8] P. Batfalsky, V.A.C. Haanappel, J. Malzbender, N.H. Menzler, V. Shemet, I.C. Vinke and R.W. Steinbrech, Chemical interaction between glass–ceramic sealants and interconnect steels in SOFC stacks, Journal of Power Sources, 2006, 155, 128-137
[9] V.A.C. Haanappel, V. Shemet, S.M. Gross, Th. Koppitz, N.H. Menzler, M. Zahid and W.J. Quadakkers, Behaviour of various glass–ceramic sealants with ferritic steels under simulated SOFC stack conditions, Journal of Power Sources, 2005, 150, 86–100
[10] A.C. Haanappel, V. Shemet, I.C. Vinke, S.M. Gross, Th. Koppitz, N.H. Menzler, M. Zahid and W. J. Quaddakkers, Evaluation of the suitability of various glass sealant—alloy combinations under SOFC stack conditions, J. Mat. Sci., 2005,40, 1583-1592

ACKNOWLEDGEMENT
    The authors would like to acknowledge the financial assistance received from the BMWi and the company Staxera (Dresden, Germany).

DETERMINATION OF FRACTURE STRENGTH OF GLASS-CERAMIC SEALANT USED IN SOFC

Kais Hbaieb
Institute of materials research and engineering (IMRE)
3 research link
Singapore 117602

ABSTRACT

The sealant is an integral component of SOFC and its integrity is very crucial for the functionality of the cell. Any damage or failure of this component may result in total operational failure of the fuel cell system. The measurement of the fracture strength of the sealant is needed in the design against failure of this component. A controlled buckling technique is used to determine the fracture strength of the sealant by axially compressing a thin ceramic substrate supporting the sealing glass-ceramic film. A simple set-up is made to measure the applied compressive displacement at failure. Using the quantity measured during the experiment, the curvature and film strength are determined using buckling theory. The fracture toughness is estimated to be $\sim 3.07 MPa\sqrt{m}$.

INTRODUCTION

Seals for solid oxide fuel cell (SOFC) must meet several requirements. Seals have to essentially be hermetic in order to prevent mixing and cross-over of the fuel and oxidants. Since they are exposed to both cathode and anode environments, they must be stable in both oxidizing and reducing atmospheres. To minimize the thermal-mechanical stresses, sealants have to possess coefficient of thermal expansion (CTE) close to other fuel/stack components. In addition, sealants must no undergo any reaction with other cell components. Sealants have to also be stable under frequent thermal cycling, preferable under high cooling and heating rates to allow for rapid start-up and shut down modes.

Glass-ceramics are preferred materials for rigid sealing concepts for SOFC. They provide sufficient gas leak-tightness for intermediate lifetime (1000 hrs) under operation conditions. They have superior mechanical properties and higher viscosity at the SOFC operating temperature than glasses. Their CTE can be matched to other cell/stack components by controlling the volume fraction of the crystalline phase formed during sintering.

However, glass-ceramic can also be vulnerable to damage and failure due to their inherent brittleness. Any crack in the seal will cause leakage and leads to lower cell performance and efficiency and poor fuel utilization. For successful design of SOFC system the mechanical properties of glass-ceramics must be assed especially fracture resistance. The aim of this paper is to measure fracture toughness of glass-ceramic using buckling testing. The limitation of this technique will be discussed.

THEORETICAL BACKGROUND

The controlled buckling test was first proposed by Chen and co-workers [1-3] to measure fracture toughness of brittle film deposited on compliant substrate. The film is very thin compared to the substrate and the sample is loaded such that the film is subject to tensile stress. As the fracture toughness of the film is very low, buckling cause the film to undergo channeling cracking. The energy release rate for cracking of thin film under residual tension can be theoretically estimated as described by Beuth and summarized below [4].

Beuth [4] provided solution for the fracture mechanics quantities as function of the properties of the film/substrate system. Two main problems were analyzed: a plane strain 2-D crack in a film oriented perpendicular to the film/substrate interface and that is either partly or fully extending across the film thickness. In the latter case the crack tip touches the interface and when the crack channels the film along the interface the energy release rate quickly reaches a steady-state. A 2D plane strain analysis is sufficient to quantify the steady-state energy release rate:

$$G_{ss} = \frac{1}{2}\frac{\sigma_f^{\,2}}{\bar{E}_f} t_f \pi g(\alpha, \beta) \tag{1}$$

where σ_f, t_f are the applied stress and the thickness of the film, respectively. The quantities α and β are the Dundurs parameters expressing the elastic mismatch between film and substrate. For plane strain problems α and β are given by:

$$\alpha = \frac{\bar{E}_f - \bar{E}_s}{\bar{E}_f + \bar{E}_s}, \qquad \beta = \frac{\bar{E}_f\left(\dfrac{1-2\nu_s}{1-\nu_s}\right) - \bar{E}_s\left(\dfrac{1-2\nu_f}{1-\nu_f}\right)}{2\left(\bar{E}_f + \bar{E}_s\right)} \tag{2}$$

where $\bar{E}_f, \bar{E}_s, \nu_f, \nu_s$ are the plane strain elastic moduli and the Poisson's ratios of the film and substrate, respectively. The factor g in equation (1) depends on the Dundurs parameters and was plotted and tabled for the full range of α in [4].

For the partially cracked film, two modes of crack extensions can take place. The 2-D plane strain crack can extend perpendicular to the interface and 3-D crack can channel through the film and along the interface. It is important to note that the plane strain cracking can take place with or without channeling cracking. The expression for the energy release rate for channeling cracking is analogous to the fully cracked film described above but now depends also on the crack length within the film:

$$G_{ss} = \frac{1}{2}\frac{\sigma_f^{\,2}}{\bar{E}_f} t_f \pi g\left(\alpha, \beta, \frac{a}{t_f}\right) \tag{3}$$

Where a is the crack length. The factor g is plotted in [4]. A normalized stress intensity factor is also plotted. It is noted that for compliant film (relative to the substrate) the normalized stress intensity factor is nearly constant. When referring to Fig. 7 in [4] for the normalized stress intensity factor, we can approximately estimate the energy release rate for 2D plane strain crack (partially cracking the film) as:

$$G_p \approx 0.64\frac{\sigma_f^{\,2}\pi t_f}{\bar{E}_f} \tag{4}$$

Beuth further discussed the mode of cracking for partly cracked film. If the fracture toughness of the film is higher than the energy release rate for plane strain 2-D crack then no cracking takes place. Likewise, if the fracture toughness of the film is larger than the steady-state energy release rate for channeling cracking then no channeling cracking can take place. For the case, of $\alpha=-0.8$ and $\beta=-\alpha/4$, Beuth mapped out the region (Fig.11 in [4]) where 'No cracking', 'Plane strain cracking & No channeling' and 'Plane strain cracking & channeling' can take place.

EXPERIMENTAL METHOD
Sample preparation

Borosilicate based glass-ceramic was screen printed on both sides of 3 mol% yttria partially stabilized zirconia (3YSZ) substrates. These substrates are either thin substrates of 300±30 μm thickness supplied by Kerafol GmbH, Germany or of 500 μm thickness (hot isopressed -HIPed) supplied by CoorsTek, USA. The substrates are 50mm square plates and the printed area is 40mm square. The samples were subsequently sintered in air at 1050°C for 2 hours. After sintering the samples were cut into thin parallel strips of ~3mm and tested under buckling load.

Testing procedure

The controlled buckling testing technique is used for the measurement. Two special set-ups were made for the test (Fig.1). The sample is placed between two clumps. One of the clumps is fixed, while to other is movable. In one of the set-ups (Fig. 1a) a micrometer is attached to the movable clump and compressive load is applied by rotating its spindle. The micrometer has thus the role of loading the sample and measuring the contraction. This design is limited to very thin samples and can not buckle the HIPed substrate of 0.5mm thickness. A second design was made and the loading is performed using a fixed plate-guided screw that drives a second movable clump/plate to contract the sample. The micrometer is separately attached to the outer fixed plate through which the screw is driven (Fig. 1b). Before the test the micrometer is zeroed and after the sample is brought to failure the compressive displacement is measured. The movable plate is tightly enclosing a symmetric trapezoidal base to ensure no tilting/shifting while loading.

Fig. 1: Two set-ups used for the buckling test; a) micrometer is used to load the sample and simultaneously measure the contraction, b) load is applied through a screw driving a movable plate contacting the sample, a micrometer is attached to a fixed plate to measure the movement of the loading plate.

Fig. 2 shows the schematic illustration of the sample after buckling and the important geometric dimensions required for fracture strength estimation. The length of the sample is measured before the test. The only measurable quantity in the test is the contraction. For accurate measurement a micrometer is used as it is accommodated in both set-ups shown in Fig. 1.

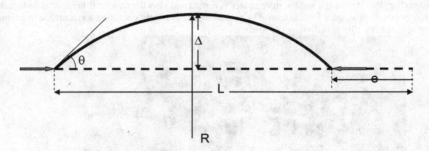

Fig. 2: Schematic representation of the sample under buckling loading. Important geometrical quantities are shown.

Since it is very slender, the sample buckles after application of small compression strain. If the sample fails under channeling cracking, the loading must be stopped immediately after the first crack appears, otherwise the fracture toughness will be overestimated. However, in our case and in all instances, no channeling cracks are observed and the total sample including both films and substrate cracked catastrophically. Thus, the tests can be carried out until the fracture point and there is no need to slowly loading the sample to spot the first appearance of channeling cracking.

Since the buckling of the sample is large, large deformation buckling theory of beam is used:

$$\frac{L}{R} = 4kK(k), \qquad \lambda = 2\left[1 - \frac{E(k)}{K(k)}\right] \qquad (5)$$

Where L is the length of the specimen, $k = \sin(\theta/2)$, $K(k)$ and $E(k)$ are complete elliptic integrals of first and second and λ is the contraction ratio.

Since the film is deposited on both sides of the substrate, the neutral axis is the central line of the composite and the average strain in the film is:

$$\overline{\varepsilon}_f = \frac{(t_s + t_f)}{2R} \qquad (6)$$

A Fortran program has been written to calculate the elliptic integrals and therefore the curvature, strain and applied stress. The energy release rate can be then estimated and is reported in the next section.

RESULTS AND DISCUSSIONS

From equations (1, 3-4) the energy release rate can be calculated if film thickness, elastic modulus and stress in the film are determined. The substrate and total thickness of the sample are measured using a micrometer. The film thickness is calculated by subtracting the substrate thickness from the total thickness.

The elastic modulus is measured using nanoindentation. To prepare for the nanoindentation test, the glass-ceramic is polished down to 0.1 µm. The continuous stiffness method (CSM) is used. The indenter is loaded up to 0.5, 1, 2, 3 and 5 µm. Since the nanoindentation is very localized, 10 tests have been conducted each time. The results are given in Table 1. These results are the average over an extended range of displacement depths into the sample. Fig. 3 shows a typical example of the elastic modulus versus the depths into the samples up to 0.5 µm. The average value for the elastic modulus across the 4 sets of tests is ~81.5GPa.

Table 1: Elastic modulus measured by nanoindentation into glass-ceramic at different displacement depths.

Depth/Specimen	E(Glass-ceramic)-GPa
0.5µm	88.4±12.65
1.0µm	84.5±11.9
2.0µm	78.4±19
3.0µm	74.7±20.2
5.0µm	81.3±8.9
Average	81.5

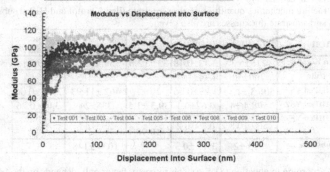

Fig. 3: Elastic modulus versus indenter displacement into the sample for the nanoindentation experiment using the continuous stiffness method and for maximum depth into the sample of 0.5 µm.

A slight residual compressive stress is developed in the glass-ceramic film upon cooling to room temperature due to mismatch in the coefficient of thermal expansions (CTE) of both the film and Kerafol substrate. This residual stress is given by:

$$\sigma^R = \frac{\Delta\alpha\Delta T E_f}{(1-v_f)\left[1+\dfrac{t_f}{t_s}\dfrac{E_f}{E_s}\right]} \tag{7}$$

Where $\Delta\alpha$ is the thermal expansion coefficient (CTE) mismatch between film and Kerafol substrate, ΔT is the temperature difference between the stress-free and room temperatures, v_f is the Poisson's ratio of the film and E_f and E_s are the elastic moduli of the film and substrates, respectively. The CTE of Kerafol substrate is $11x10^{-6}$ 1/K, and that of CoorsTek substrate is $10.3x10^{-6}$ 1/K and that of glass-ceramic film is $10.2x10^{-6}$ 1/K. Thus, there is hardly any residual stress between glass-ceramic and CoorsTek substrate upon cooling to room temperature. Note that for calculating the energy release rate, the total stress, including the biaxial residual stress and plane strain applied stress, has to be considered. The elastic modulii of both substrates are taken to be 200GPa. The Poisson's ratios of film and substrates are taken to be 0.2 and 0.23, respectively.

Table 2 summarizes the results for the fracture strength and toughness of glass-ceramic bonded to either Kerafol or CoorsTek substrates. Along with the fracture quantities, contraction and film and substrate thicknesses are also given. The same tests were also carried out on bare substrates and the fracture strain and stress are determined. As can be seen, in most cases, the sample is strained so much that the stress sustained by the substrate fall in the critical range of substrate fracture strength. Thus, it is likely that the substrate has cracked first and caused in turn the film to crack. However, in other cases the strain was not as high (for the case where the film thickness is 39μm and the contraction is 0.67mm) and the sample has nevertheless cracked. As mentioned previously no channeling cracking have been observed. Thus, for the case where the sample has cracked at a strain lower than the substrate fracture strain, equation (4) could be used to estimate the fracture toughness of the film. The stress intensity factor in this case is as follows:

$$K_c = 1.418x10^{-3}\sigma_f\sqrt{t_f} \qquad (8)$$

where the film thickness is in microns. Equation (8) is used to determine the critical stress intensity factor given in Table 2 and was estimated to be $\sim 3.07 MPa\sqrt{m}$.

Table 2: Fracture mechanics quantities of glass-ceramic film, Kerafol and HIPed substrates. Contraction, film and substrate thicknesses are also given.

Specimen	t_s (μm)	e (mm)	t_f (μm)	Fracture stress (MPa)	K_{IC} (MPam$^{1/2}$)
Glass-ceramic on Kerafol	215±5.3	3.43±0.48	33.1±2.6	359.5±22.1	2.26±0.14
Kerafol substrate	200±30	4.34±1.88	-	883±185	-
Glass-ceramic on HIPed YSZ	506.3 ±1.15	1.29±0.06	20 ±1	462 ±11.93	2.93 ±0.06
Glass-ceramic on HIPed YSZ	508.4 ±4	0.67±0.1	39.3 ±1.5	338±24	3.07±0.26
Glass-ceramic on HIPed YSZ	509 ±0.8	0.79±0.2	59.2 ±0.96	395±47	-
HIPed YSZ Substrate	513 ±2	1.17 ±0.3	-	1094 ±148	-

Note that, according to equation (8), the stress intensity factor only depends on the film thickness and independent of crack length. This is the case for a thin film where the initial flaw is very small. When the film thickness becomes very large the film behaves similar to a bulk sample and the crack initiates from a flaw within the film and catastrophically propagates, fracturing the entire sample. In this case, it is not possible to estimate the fracture toughness. In table 2, no fracture toughness is given for the film with a thickness of ~59μm.

It is difficult to observe channeling cracks if the substrate and film are strained. A good alternative to the current procedure is to force the film to strain without loading the substrate. This could be achieved by choosing a substrate with much lower CTE than that of glass-ceramic. The fracture stress can be written as function of the critical stress intensity factor for channeling cracking as follows:

$$\sigma_f^c = \frac{10^3 K_c}{1.253 \sqrt{g t_f (\mu m)}} \tag{9}$$

If we take the fracture toughness of glass-ceramic to be $3.07 MPa\sqrt{m}$ and film thickness to fall in the range of $20 - 25 \mu m$, then the fracture stress will be in the range of $480\text{-}535 MPa$. Using equation (7), this corresponds to a CTE mismatch of $4.9\text{-}5.5 \times 10^{-6}$ 1/K. Thus, a substrate made of mullite with a CTE of $\sim 5.2 \times 10^{-6}$ 1/K could be a potential good substrate to use to trigger channeling cracking in glass-ceramic.

CONCLUSIONS

Controlled buckling testing technique was used to determine the fracture toughness of screen-printed glass-ceramic film. In some cases, the bending stress was so high that substrate was likely cracked before the film. In other cases, the buckling strain was lower that the substrate fracture strain and the plane strain cracking was concluded to be the mode of fracture of glass-ceramic film. Although, no channeling was observed it is proposed that using mullite could be an alternative to induce channeling cracking in glass-ceramic. The fracture toughness was estimated to be $\sim 3.07 MPa\sqrt{m}$.

REFERENCES

[1] Chen, Z., Cotterell, B. and Wang, W., 2002, "The fracture of brittle thin films on compliant substrates in flexible displays," Engng Fract. Mech., 69, pp. 597-603.
[2] Balakrisnan, B., Chum, C. C., Li, M., Chen, Z. and Cahyadi, T., 2003, "Fracture toughness of Cu-Sn intermetallic thin films," J. Elec. Mater., 32, pp. 166-171.
[3] Chen, Z. and Gan, Z., 2007, "Fracture toughness measurement of thin films on compliant substrate using controlled buckling test," Thin Solid Films, 515, pp. 3305-3309.
[4] Beuth, J. L., 1992, "Cracking of thin bonded films in residual tension," Int. J. Solids struct., 29, pp. 1657-1675.

CREEP BEHAVIOR OF GLASS/CERAMIC SEALANT USED IN SOLID OXIDE FUEL CELLS

W.N. Liu, X. Sun, B. Koeppel, and M.A. Khaleel
Pacific Northwest National Laboratory
Richland, WA 99354

ABSTRACT

Creep deformation of glass sealant used in Solid Oxide Fuel Cell (SOFC) under operating temperature is not negligible. The goal of the study is to develop a creep model to capture the creep behavior of glass ceramic materials at high temperature and to investigate the effect of creep of glass ceramic sealant materials on stresses in glass seal and on the various interfaces of glass seal with other layers. The creep models were incorporated into SOFC-MP and Mentat FC, and finite element analyses were performed to quantify the stresses in various parts. The stress in glass seals were released due to its creep behavior during the operating environments.

INTRODUCTION

Among various SOFC designs, anode-supported planar cells have shown great potential in delivering high performance at reasonable costs [1, 2]. Planar SOFCs offer a significant advantage of a compact design along with higher power densities. In the meantime, they require the incorporation of hermetic gas seals for efficient and effective channeling of fuel and oxygen.

Seals should function in the high operating temperature of SOFC and at the complicated operating environments [3-5] to prevent the leakage of air and fuel, to isolate the fuel from the oxidant, and to insulate the cell from short circuit. Glass joining, as rigid seal, provides a cost effective and relatively simple method of bonding ceramic and metal parts. However the softening point of the glass component typically limits the maximum operating temperature to which the joint may be exposed. The glass/ceramic sealant materials developed by Pacific Northwest National Laboratory (PNNL) possess relatively high stiffness even at the high operating temperature of SOFC due to presence of the ceramic phase [6, 7]. However, creep deformation of glass/ceramic sealant under the typical cell operating temperature should not be neglected. The temperature differential between the initial, stress-free fabrication temperature and SOFC operating temperature will cause stresses and deformation of various cell components. Under long term operation, the stresses and deformation are expected to relax and re-distribute due to the anticipated creep behavior of the glass sealant and interconnect material.

In this paper, the creep behavior of the glass/ceramic sealant materials is experimentally measured. The creep behavior of the glass/ceramic sealant is modeled and the material parameters used for the creep model are calibrated. The goal of the study is to investigate the long term performance of the fuel cell stack. The temperature differential between the initial, stress-free fabrication temperature and SOFC operating temperature will cause stresses and deformation of various cell components. It is found that creep behavior of the ferritic stainless steel IC contributes to narrowing of both the fuel and the air flow channels. Stresses relaxed in IC due to creep will be transferred to other parts and cause the re-distribution of stresses in SOFC.

EXPERIMENTAL MEASUREMENT OF CREEP BEHAVIOR OF GLASS/CERAMIC SEALANTS

Creep test is implemented in the temperature chamber at the different temperatures. Loading case of the test is illustrated in Figure 1. The compressive stress is applied to the cylinder samples. Short-term compression creep tests were performed at various temperatures (700, 750, and 800 °C) and load levels to evaluate the creep strain with time. These materials did exhibit a creep rate that increased with temperature and applied stress. The glass/ceramic samples amounted in the load case as shown in Figure 1 are placed in the temperature chamber and heated to the prescribed temperature first, then the compressive load is applied to the glass/ceramic sample by the frame within 10 min and kept constant for about 2.5 h. The sample used in the test is cylinder with 10 mm height and 5 mm diameter. The cross-head movement is recorded to calculate the deformation of the glass/ceramic samples. The cross-head movement consists of two parts: the deformation of the glass/ceramic samples and the metal frame applying load on the sample.

The creep strains measured at 800 °C are depicted in Figure 2 for three different loading stresses, i.e., 10 MPa, 20 MPa, and 30 MPa. The initial elastic deformation corresponding to the lowest load 10 MPa was excluded in the curves.

Figure 1 Load case used in the creep test

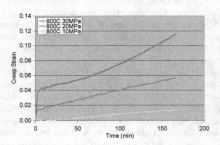

Figure 2 Measured creep strains at 800 °C for three different loading stresses

CREEP MODEL OF GLASS/CERAMIC SEALANTS

The glass ceramic seal used in this study was developed by PNNL for planar-type SOFC applications, which is a barium–calcium–aluminosilicate (BCAS)-based glass with the addition of boron oxide. It includes two distinct phases: amorphous glass phase and crystalline ceramic phase. To model its creep behavior under the high operating temperature of SOFC, a two phases model is shown in Figure 3, where glass phase is characterized by creep behavior, and the crystalline ceramic phase is no creep.

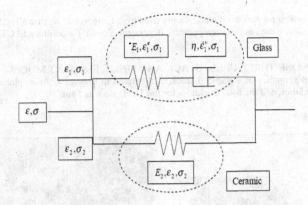

Figure 3 Two phase model for glass/ceramic sealant materials

The stress and strain for each element in Figure 3 should satisfy the following relationship as for glass phase,

$$\sigma_1 = E_1 \varepsilon_1^e = \eta \dot{\varepsilon}_1^v \tag{1a}$$

and for ceramic phase

$$\sigma_2 = E_2 \varepsilon_2 \tag{1b}$$

where σ and ε represent stress and strain. The resultant stress and strain satisfy

$$\sigma = \sigma_1 + \sigma_2 \tag{2a}$$
$$\varepsilon = \varepsilon_1 = \varepsilon_2 \tag{2b}$$

where

$$\varepsilon_1 = \varepsilon_1^e + \varepsilon_1^v \tag{3}$$

Visco-elastic model is adapted to describe the creep behavior of the glass ceramic seal materials. The creep rate may be determined as

$$\dot{\varepsilon}_c = \frac{1}{E_1 + E_2} \frac{E_1}{\eta} \left(\frac{\eta}{E_1} \dot{\sigma} + \sigma - E_2 \varepsilon \right) \tag{4}$$

where subscripts "1" and "2" refer to the glass phase and ceramic phase, respectively. E represents modulus and η denotes viscosity of the glass phase. These mechanical properties are calibrated by the creep test at the different temperature and different loading forces.

The mechanical properties of a commonly used high temperature ferritic alloy SS430 is used in the current numerical analyses. The temperature dependent Young's modulus and CTE for SS430 are used as given in [8].

GEOMETRY OF ONE/THREE CELLS STACK AND FINITE ELEMENT MODEL

The initial geometry of a one-cell stack is illustrated in Figures 4. Its in-plane dimensions are 157.9mm by 149.5mm, and the total thickness for the 1-cell stack is 5mm.

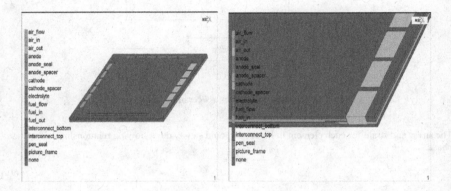

Figure 4 Illustration of 1-cell stack

Since the SOFC components are considered to be stress free at its assembly temperature of 800°C, different degrees of thermal stresses will be generated at the steady state operating temperature due to the CTE mismatch of various cell components. Figure 5 shows the finite element model used in the thermal-mechanical analysis.

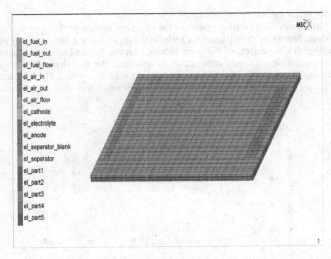

Figure 5 Finite element model used in the thermal-mechanical analysis

To obtain the steady-state operating temperature profile, SOFC-MP was used to perform the electro-chemical-thermal analyses. Figure 6 depicts the steady state IC temperature distributions on the surfaces.

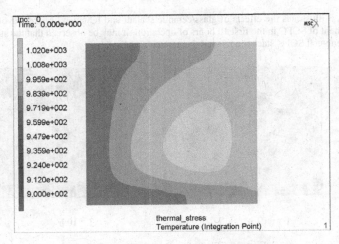

Figure 6 Steady state temperature distributions

NUMERICAL RESULTS AND DISCUSSIONS

Figure 7 illustrates the modeling results of the geometry change of fuel and air flow channel as well the stress/strain history in PEN (Positive-Electroltyte-Negative) seal during steady state operation condition. It is clear that the effects of the creep behavior on the fuel and air flow channel heights are different. During the long-term steady state operation condition, the fuel channel is narrowing, but air channel is widening. Reduction of the fuel channel height reaches up to 2.4% after the steady state operation of 10 hour, and the curve shows us that the narrowing of fuel channel will not stop but just slow down.

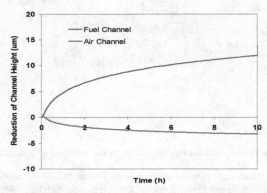

Figure 7 Effect of creep behavior on geometry stability of the fuel and air flow channels

Figure 8 illustrates the effects of glass/ceramic sealant and IC creep on stress evolution in the anode component of SOFC in the first 10 hours of operation. It may be observed that the stresses in the anode of the one-cell SOFC stack are relaxed due to creep.

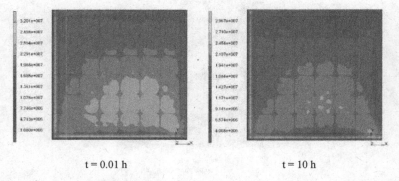

t = 0.01 h   t = 10 h

Figure 8 Effect of creep on stress in anode

CONCLUSIONS

The effects of glass/ceramic sealant and interconnect creep on the change of air and fuel flow channels were predicted under the steady state operating condition for a period of 10 hours. The purpose is to quantify the long term SOFC stack performance by investigating the changes in air and fuel flow paths and the corresponding stress re-distribution among various cell components. The conclusions of the current study can be summarized as follows. Over the short term course of 10 hours of operating time, creep behavior of the glass/ceramic sealant and interconnect contributes to the height reductions of the fuel flow channels. Creep of the glass/ceramic sealant and interconnect reduces the stress level in the PEN structures and redistribute the stresses.

ACKNOWLEDGEMENTS

The Pacific Northwest National Laboratory is operated by Battelle Memorial Institute for the United States Department of Energy under Contract DE-AC06-76RL01830. The work was funded as part of the Solid-State Energy Conversion Alliance (SECA) Core Technology Program by the U.S. Department of Energy's National Energy Technology Laboratory (NETL).

REFERENCE

[1]W. P. Teagan, J. H. J. S. Thijssen, E. J. Carlson, and C. J. Read, Current and future cost structures of fuel cell technology alternatives, *Proc. 4th European Solid Oxide Fuel Cell Forum*, A.J. McEvoy (Ed.), Lucerne, Switzerland, 2, 2000, pp. 969-980.

[2]B. W. Chung, C. N. Chervin, J. J. Haslam, A. Pham, R. S. Glass, Development and characterization of a high performance thin-film planar SOFC stack, *Journal of the Electrochemical Society*, 152 (2), 2005, pp. A265-269.

[3]K. A. Nielsen, M. Solvang, S. B. L. Nielsen, A. R. Dinesen, D. Beeaff, P. H. Larsen, Glass composite seals for SOFC application, *Journal of the European Ceramic Society*, 27 (2-3), 2007, pp.1817-22.

[4]S. R. Choi, N.P. Bansal, Mechanical, Mechanical properties of SOFC seal glass composites, *Ceramic Engineering and Science Proceedings*, 26 (4), 2005, pp. 275-83.

[5]F. Smeacetto, M. Salvo, M. Ferraris, J. Cho, A.R. Boccaccini, Glass-ceramic seal to join Crofer 22 APU alloy to YSZ ceramic in planar SOFCs, *Journal of the European Ceramic Society*, 28 (1), 2008, pp. 61-8.

[6]K. S. Weil, J. E. Deibler, J. S. Hardy, D. S. Kim, G. G. Xia, L. A. Chick, and C. A. Coyle, Rupture testing as a tool for developing planar solid oxide fuel cell seals, *Journal of Materials Engineering and Performance*, 13 (3), 2004, pp.316-326

[7]W.N. Liu, X. Sun, M.A. Khaleel, Predicting Young's modulus of glass/ceramic sealant for solid oxide fuel cell considering the combined effects of aging, micro-voids and self-healing, *Journal of Power Sources*, v 185, n 2, 1 Dec. 2008, p 1193-200

[8]K. I. Johnson, V. N. Korolev, B. J. Koeppel, K. P. Recknagle, M. A. Khaleel, D. Malcolm and Z. Pursell, Finite Element Analysis of Solid Oxide Fuel Cells Using SOFC-MP™ and MSC.Marc/Mentat-FC™, PNNL report 15154, Pacific Northwest National Laboratory, June 2005.

THERMAL CYCLE DURABILITY OF GLASS/CERAMIC COMPOSITE GAS-TIGHT SEALS ON METAL SUBSTRATES

Seiichi Suda, Masahiko Matsumiya, Koichi Kawahara and Kaori Jono
Japan Fine Ceramics Center
Nagoya, Aichi, Japan

ABSTRACT

Small-sized SOFC is required to endure prompt startup and shutdown operations as well as efficient energy generation to apply it to small electric power generators of handy devices and vehicles, but conventional glass seals are generally easy to break by prompt heating or cooling. We have thus developed glass/ceramic sealing materials to use as gas-tight seals of portable SOFC. The composite seals was composed of glass precursor particles (NS particle) and amorphous silicate particles (S particles) working as a filler. Thermal cycle durability was investigated for the NS/S composite seals after the seals were fused on SUS430 and Ag plates as well as porous LSCF ceramics. The seals fused on SUS430 and Ag plates exhibited different thermal cycle profiles from those on porous LSCF ceramics. SEM observation of the seals before and after the thermal cycle indicated that the cycle brought about aggregation of pores in the seals and the grown pores by the cycle would contribute the expansion of cracks. The formation of cracks that brought about severe gas leakage was caused by thermal stress, but the composite seals partly released thermal stress at glass soften temperature of 530°C. Softening profiles of the glass component in the composite seals would be important for developing thermal cycle durable gas-tight seals.

INTRODUCTION

Small-sized solid oxide fuel cell (SOFC) is a promising energy generation system because it can generate high electric power density in a small occupied space. Small SOFC will be widespread as portable fuel cell systems if it generates high electronic power with consuming natural gas or reformed fuel gas efficiently at relative low temperatures. Portable SOFC can be applied as small electric power generator for handy devices and vehicles. Portable SOFC are required prompt startup and shutdown operations. Development of small SOFC that can endure repeated prompt startup and shutdown remains some issues. As for the durability of cell, small tubular cells were used because the tubular cells actually led to improve thermal cycle durability[1].

Gas-tight seals will be another issue for the development of portable SOFC. Gas-tight seals are indispensable for fabricating high-performance portable SOFC and gas leakage from seals would bring about severe damage to cells or stacks. Gas-tight sealing materials are required to exhibit low fusion temperatures to avoid excess sintering of SOFC components and not to remain excess internal stress induced during the fusion. The seals also should possess long-term stability at working temperatures both in air and a reduction atmosphere because the sealing materials for SOFC directly contact with either air or a fuel gas. Frequent startup and shutdown operation will give us another challenge for development of highly reliable gas-tight seals. For example, glass seals exhibit tight adhesion with dense ceramics and relatively stable gas-shielding properties at working temperature, but they generally lack thermal cycle durability.

We have thus developed glass/ceramic sealing materials to apply to gas-tight seals for portable SOFC. Two kinds of silicate precursor particles were synthesized by sol-gel processes. One of the silicate particles is sodium-containing particles (NS particles) and the other is sodium-free silicate particles (S particles). NS particles will result in glass after the fusion at 800-850°C and S particles would be act as fillers[2] (Fig. 1). NS particles and S particles were mixed by ball milling and flexible green sheet was synthesized with both the particles and some organic binders by tape-casting method. Insufficient fusion of NS particles brought about degradation with steam at high temperature, but an appropriate fusion of the particles mixed with NS and S particles resulted in glass/ceramic seals that exhibited sufficient gas shield properties in moist hydrogen at 700°C[3]. An adequate ratio of S particles to NS particles also improved adhesion with porous SOFC cathode materials[4]. When the glass/ceramic composite seals containing small amount of S particles was fused on porous cathode materials, excess amount of the fused fluid was penetrated into the porous materials and the seals exhibited insufficient sealing properties. On the other hand, excess addition of S particles brought about insufficient adhesion strength. Static gas-shielding properties of the composite sealing materials were sufficient high and these properties were almost independent of the ratio of S particles to NS particles (NS/S ratio). Thermal cycle durability of the seals, however, much depended on NS/S ratio[5]. When the seal was fused with the porous SOFC cathode material of $La_{0.6}Sr_{0.4}Co_{0.2}Fe_{0.8}O_3$ (LSCF), the appropriate composite seals with NS/S ratio of 80/20 showed highly thermal cycle durability and no gas leakage was observed in more than several months. The composite seals can be sufficiently applied on the fusion with porous LSCF and dense electrolyte.

SOFC system contains not only cell components made of porous or dense ceramics but also manifolds and current collectors that are required gas-tight seals. Lowering SOFC working temperature actually led to use metallic manifolds or separators, but gas-tight seals between cell components and some conventional metals will remain some challenges. Heated and partly softened seals may be expected to work as relaxation of stress induced by thermal expansion mismatch in the case where thermal expansion of conventional metals was much larger than that of ceramic cell components. We then investigated thermal cycle durability of the composite seals fused with two kinds of metals of SUS430 and Ag. Increasing and decreasing temperature rate was changed from 250°C/h to 450°C/h and thermal cycle between 400°C and 650°C was repeated in air. Thermal expansion profiles were also investigated to discuss the thermal expansion mismatch.

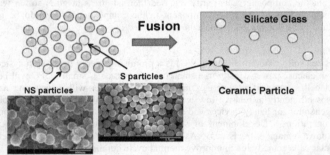

Fig. 1.   Schematic illustration of glass/ceramic composite seals by fusion with NS and S particles.

EXPERIMENTAL PROCEDURE

Both NS particles and S particles were synthesized by sol-gel processes. NS particles were synthesized by sol-gel and simultaneous ion-exchange method. Hydrolysis of alkoxide generates OH groups and this synthesis method is promoted ion-exchange between alkali ions and protons of OH groups just before condensation. The process resulted in alkali-containing mono-dispersed spherical particles[6]. NS particles were synthesized with concentrated NaOH solution and tetraethyl orthosilicate (TEOS) at the NaOH/TEOS molar ratio of 0.8. TEOS, hydroxyl propylcellulose and concentrated NaOH solution was mixed with ethanol and stirred vigorously at 50-60°C in air. Stirring for 1 h resulted in suspension dispersed with NS particles. Na/Si ratio of NS particles was estimated to be 0.63 from the results by energy-dispersive X-ray spectroscopy (EDS). S particles were synthesized by a similar method to the preparation of NS particles. The preparation of S particles was used NH_3 solution instead of the concentrated NaOH solutions. Obtained NS particles were put into de-ionized water to promote hydrolysis of remaining alkoxide and calcined at 300°C for 12h in air. The calcined NS particles and S particles were mixed at the mass ratio of NS/S=100/0(only NS particles), 95/5, 90/10, 85/15, 80/20, 75/25 and 70/30, and flexible glass/ceramic precursor green sheets with a thickness of 50-250 μm were obtained with the mixed particles and some organic binders by tape casting method.

Thermal cycle durability was estimated using the seal testing equipment illustrated in Fig. 2 after the fusion with porous LSCF ceramic, SUS430 plate or Ag plate. The flexible NS/S sheets were attached on these substrates and fused at 800°C for 2 h in air. Gas leakage was estimated as change in vacuum pressure with time. Vacuum small chambers that were closed with fused seals were set in furnace. The temperature of the furnace was increased and decreased at the rate of 250°C/h to 450°C/h. When thermal cycle durability was previously investigated for conventional soda-lime glasses, cracks in glasses or separation between glasses and some substrates were often formed on cooling at several tens temperature lower than glass transition temperature. The glass composition derived from NS particles showed glass transition temperature of 470°C, and thermal cycle was thus repeated between 400°C and 650°C for 120 h. Vacuum pressure and temperature of the chamber were monitored every ten seconds.

Interfaces between substrates and the composite seals were observed with an optical microscope and a scanning electron microscope (SEM). Thermal expansion profiles were also investigated at temperatures from room temperature to 760°C. Increasing temperature rate was set to be 5°C min⁻¹, and thermal expansion profiles of SUS430 was measured in nitrogen and those of other substrate and the composite seals were measured in air.

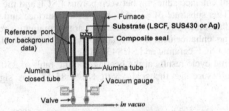

Fig. 2. Schematic illustration of gas seal tester.

RESULTS AND DISCUSSION

Table 1 shows thermal cycle durability for the NS/S composite seals fused on porous LSCF. The glass derived from only NS particles (NS/S=100/0) showed poor durability, but the addition of S particle to NS particles improved the durability. Excess addition of S particles also diminished the durability because S particles hardly contributed the adhesion between the composite seal and porous LSCF, and fusion of the composite seals with NS/S=70/30 would bring about poor adhesion strength as compared to other composite seals. The composite seal with NS/S=80/20 exhibit highly thermal cycle durability for more than 3000h, which is more than 370 cycles, and an adequate composition ratio resulted in highly durable gas-tight seals.

The composite seals were then fused with SUS430 plate instead of porous LSCF. SUS430 materials are generally used as metal manifolds and some gas pipes. Fig. 3 shows thermal cycle profiles of various NS/S composite seals fused on SUS430 plate. Heating and cooling rate was set to be 250°C/h and the thermal cycle between 400°C and 650°C was repeated 40 times. Vacuum pressure was steeply decreased for the seal with NS/S=70/30. The composite seals of NS/S=70/30 showed

Table 1.　Relationship between heating/cooling rate and thermal cycle durability for the composite seals fused on porous LSCF.

NS/S	100/0	95/5	90/10	85/15	80/20	75/25	70/30
Rate: 250°C/h	NG	NG	Good	Good	Good*	Good	NG
Rate: 350°C/h			NG	Good	Good	Good	
Rate: 450°C/h				Good	Good	NG	

Good: Durable for >120h, Good* : for >3000h, NG: Gas leak.

poor thermal cycle durability.　Low adhesion strength owing to excess addition of S particles would bring about this low durability and it was similar to the results for the seals fused on porous LSCF. No gas leakage was observed for a long time, when the seals with NS/S=85/15 was fused on porous LSCF, but the vacuum pressure was gradually decreased for the seals fused on SUS430.　It will be more challenging to develop highly durable composite seals fused on SUS430 than those on porous LSCF.　First of all, thermal expansion coefficient of SUS430 is larger than that of porous LSCF. Remaining stress after the fusion with SUS430 would be much larger than that with porous LSCF and the large stress would be more likely to form cracks in the composite seals.　Secondly, when the composite seal was fused with porous LSCF, the fused fluid was penetrated into porous LSCF. Appropriate penetration will enhance adhesion between porous LSCF and the composite seal owing to a type of anchor effect.　This substrate of SUS430 had smooth surface and SUS430 substrate is not expected to enhance the adhesion by this anchor effect.　Large remaining stress derived from thermal expansion mismatch and no enhancement by the anchor effect would lead different thermal cycle profiles for between porous LSCF ceramic and SUS430 plate.

　　　　Fig. 4 shows thermal cycle results at the heating/cooling rates of 350°C/h and 450°C/h.　The degradation of vacuum pressure for the seals with NS/S=85/15 was significantly observed by

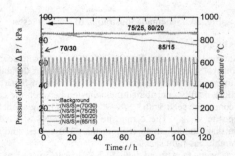

Fig. 3.　Thermal cycle durability profiles of various NS/S composite seals fused on SUS430 plate.　Heating and cooling rate was 250°C/h.

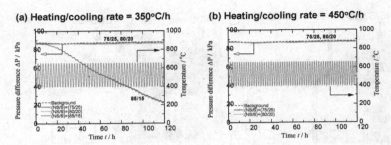

Fig. 4.   Thermal cycle durability profiles of various NS/S composite seals fused on SUS430 plate.   Heating and cooling rates were (a) 350°C/h and (b) 450°C/h.

increasing the rate.   The composite seals with NS/S=75/25 and 80/20 showed high durability under the prompt heating and cooling condition of 450°C/h.   When the seals were fused on porous LSCF, the seal with NS/S=75/25 showed slight degradation under the rate of 450°C/h, but Fig. 4(b) exhibited no degradation on the fusion with SUS430 plate.   The reason why the fusion on SUS430 showed higher durability than porous LSCF for the seal with NS/S=75/25 is still vague.   The addition of S particles would increase mechanical strength for the fused seals and the increasing mechanical strength may prevent the seals from forming large cracks that lead to severe gas leakage.   This increase in mechanical strength may improve the thermal cycle durability.   On the other hand, the durability for the seals fused on porous LSCF will be greatly influenced by contact area between the seals and porous

LSCF.   Viscosity of fusing fluid of the seals was increased with increasing ratio of S particles. Highly viscous fluid would not sufficiently infiltrate into small pores of LSCF and lead to decrease contact area between the seals and the LSCF.   The low contact area will bring about poor adhesion as shown for the seal with NS/S=75/25.

Fig. 5 shows thermal cycle profiles for the composite seals fused on Ag plate.   Only the seal with NS/S=75/25 showed no gas leakage profile for 120h, and vacuum pressure for the seal with even NS/S=80/20 was decreased with repeating thermal cycles.   The low thermal cycle durability for the

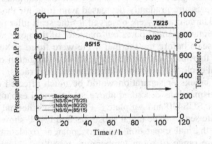

Fig. 5.   Thermal cycle durability profiles of various NS/S composite seals fused on Ag plate. Heating and cooling rate was 250°C/h.

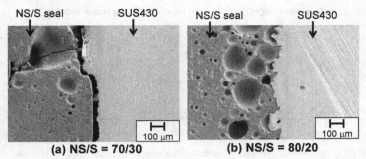

(a) NS/S = 70/30   **(b) NS/S = 80/20**

Fig. 6.   SEM images of the interface between NS/S seals and SUS430 after thermal cycle for 120h under the heating/cooling rate of 250°C/h.

seals fused on Ag plate would be due to large thermal expansion mismatch and relatively large remaining stress.   Ag plate possesses high thermal expansion coefficient of 23.1×10^{-6} K^{-1} as compared to SUS430 (20.1×10^{-6} K^{-1}) and porous LSCF (16.0×10^{-6} K^{-1}).   Large mismatch of thermal expansion induces large stress at the interface and this mismatch easily brought about the degradation.   Only the NS/S composite seal with NS/S=75/25, which was higher strength than the seals with NS/S=80/20, exhibited highly thermal cycle durability when the seal was fused on Ag plate.

Microstructures of the interface between the NS/S seals and SUS430 plate was then observed with SEM.   Fig. 6 showed SEM images of the composite seals with NS/S=70/30 and 80/20 after repeating thermal cycle for 120h at the rate of 250°C/h.   The seals with NS/S=70/30 steeply degraded vacuum pressure as shown in Fig. 3.   Severe separation between the seal and SUS430 was observed in Fig. 6 (a).   Weak adhesion between the seal and SUS430 after the fusion, and the stress induced by thermal expansion mismatch would bring about the separation.   Large crack in the seal as well as the separation in the vicinity of the interface was also observed in Fig. 6 (a).   The cracks formed by thermal expansion mismatch were extended by passing through relative large pores.   Large pores in the seals would thus assist in expanding the cracks.   On the other hand, the seal with NS/S=80/20 was well attached on SUS430 after the thermal cycle, but many voids were observed in the vicinity of the interface.

Fig. 7 shows SEM images of the interface between the seal and Ag plate before and after the thermal cycles.   Vacuum pressure gradually decreased with time for the seals with NS/S=80/20, but no apparent separation and large cracks were observed for the seals after the thermal cycles.   The seal before the thermal cycle contained small pores but diameter of the pores was relatively increased after the thermal cycles.   The seals before the thermal cycle were fused at 800°C, but this fusion did not lead to grow pores.   The thermal cycle would bring about aggregation of pores in the seals and the grown pores by the thermal cycle will contribute the expansion of cracks.

The formation of cracks or separation would be much affected on remaining stress derived from thermal expansion mismatch as described above, and thermal expansion profiles were measured for various composite seals, SUS430 and Ag plate.   Fig. 8 shows thermal expansion profiles for some substrates and the composite seals.   The metal substrates and LSCF showed almost linear relationship between temperature and thermal expansion, and no apparent oxidation or phase transitions were observed at the temperature range of room temperature to 800°C.   Instead of these substrates, the composite seals showed unique profiles.   The slopes of profiles for the seals with NS/S=80/20, 90/10 and 100/0 were changed at the glass transition temperature (T_g) of 470-480°C.   Thermal expansion

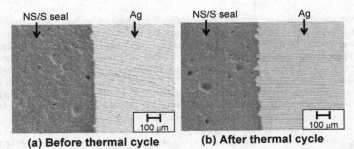

(a) Before thermal cycle   **(b) After thermal cycle**

Fig. 7.   SEM images of the interface between the composite seal with NS/S=80/20 and Ag plate (a) before and (b) after the thermal cycle for 120h under the heating/cooling rate of 250°C/h.

coefficient (CTE) at the temperature range below T_g was increased with increasing S particles in the seals.   S particles were amorphous silica particles after the fusion at 800°C and CTE of S particles was larger than that of the glass derived from NS particles.   S particles in the composite seals then increased CTE at temperature range below T_g.   The adequate addition of S particles resulted in less deviation of thermal expansions between the composite seals and metal substrates, and the less deviation would be effective to improve remaining thermal stress and thermal cycle durability on the fusion with SUS430 and Ag plate.   The profiles at the temperature range beyond Tg were mainly derived from the glass component fused with NS particles.   The composite seals were extraordinarily expanded beyond T_g and partly softened about 530°C.   The seals with NS/S=70/30 showed no apparent softened profiles, and the lack of the softened profiles would bring about the insufficient adhesion strength of the seals with NS/S=70/30.   The composite seals with NS/S=80/20, 90/10 will release remaining stress at glass soften temperature.   Softening profiles of the glass component in the composite seals would be indispensable for developing thermal cycle durable gas-tight seals.

CONCLUSION
        Thermal cycle durability was investigated for the NS/S composite seals when the seals were fused on SUS430 and Ag plates as well as porous LSCF ceramics.   The seals fused on SUS430 and

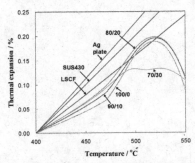

Fig. 8.   Thermal expansion profiles for various NS/S seals, SUS430 and Ag plate.

Ag plates were showed different thermal cycle profiles from those on porous LSCF ceramics. Large remaining stress owing to large thermal expansion mismatch between metal substrates and the composite seals would be more likely to form cracks in the seals. When the seals were fused on porous LSCF, appropriate penetration would enhance adhesion between porous substrate and the fused seals owing to a type of anchor effect, but no anchor effect was expected on the fusion with smooth metal substrate. The adequate addition of S particles resulted in less deviation of thermal expansions between the composite seals and metal substrates, and the less deviation would be effective to improve remaining thermal stress and thermal cycle durability on the fusion with SUS430 and Ag plate. SEM observation of the seals before and after the thermal cycle indicated that the cycle brought about aggregation of pores in the seals and the grown pores by the cycle would contribute the expansion of cracks. The formation of cracks that brought about severe gas leakage was caused by thermal stress, but the composite seals partly released thermal stress at glass soften temperature of $530°C$. Softening profiles of the glass component in the composite seals would be important for developing thermal cycle durable gas-tight seals.

ACKNOWLEDGMENTS

This work was supported by NEDO, Japan as part of the Advanced Ceramic Reactor Project.

REFERENCES
[1]Y. Fujishiro, M. Awano, T. Suzuki, T. Yamaguchi, K. Arihara, Y. Funahashi and S. Shimizu, Development of the Stacked Micro SOFC Modules using New Approaches of Ceramic Processing Technology, *ECS Trans.*, **7**, 497-501 (2007).
[2]S. Suda, K. Kawahara, K. Jono, M. Matsumiya, Development of Melting Seals Embedded Electrical Conduction Paths for Micro SOFCs, *ECS Trans.*, **12**, 283-289 (2008).
[3]M. Matsumiya, S. Suda, K. Kawahara, K. Jono, Viscoelastic behaviors of fused seals composed of glass and alloy for micro-SOFC stacking, *Proc. MRS 2008*; in received.
[4]S. Suda, K. Kawahara, K. Jono, Development of Insulating and Conductive Seals for Controlled Conduction Paths, *ECS Trans.*, **7**, 2437-2442 (2007).
[5]S. Suda, M. Matsumiya, K. Kawahara, K. Jono, Thermal Cycle Reliability of Glass/Ceramic Composite Gas Sealing Materials, Proc. Mater. Sci. & Technol. 2008, 373-379 (2008).
[6]S. Suda, T. Yoshida, K. Kanamura, T. Umegaki, Formation mechanism of amorphous Na_2O-SiO_2 spheres prepared by Sol-Gel and Ion-Exchange method, *J. Non-Cryst. Solids*, **321**, 3-9 (2003).

Mechanical Behavior

MECHANICAL PROPERTIES OF CATHODE-INTERCONNECT INTERFACES IN PLANAR SOFCs

Yanli Wang, Beth L. Armstrong, Rosa M. Trejo, Jianming Bai, Thomas R. Watkins and Edgar Lara-Curzio

Materials Science & Technology Division
Oak Ridge National Laboratory,
Oak Ridge, TN

ABSTRACT

The residual stresses in manganese cobaltite, i.e., $Mn_{1.5}Co_{1.5}O_4$, coatings applied onto alloys 441 and Crofer 22 APU were determined by X-Ray Diffraction. The residual stresses were found to be tensile at 800°C for both systems. The residual stress for spinel-coated AL441 relaxed with time and reached a value of 0.16 ± 0.02 GPa after 300 minutes. The stress relaxation process was slower for spinel-coated Crofer and reached a value of 0.23 ± 0.01 GPa after 500 minutes.

Four-point bend SENB testing technique was used to evaluate the toughness of the interfaces between LSM10, i.e., $(La_{0.9}Sr_{0.1})_{0.98}MnO_{3+\delta}$, and spinel-coated AL441 and Crofer. Sandwich test specimens were prepared by sintering the LSM10 layer at 900°C for four hours in air or under pO_2 cyclic conditions. The strain energy release rate was found to be 1.52 ± 0.11 J/m^2 for regular sintering and 1.47 ± 0.15 J/m^2 for sintering with cyclic pO_2 treatment. This difference was found to be statistically insignificant.

INTRODUCTION

Low cost chromia-forming ferritic metallic interconnects are promising candidates for Solid Oxide Fuel Cells (SOFCs) that operate up to 800 °C[1]. Presently, manganese cobaltite spinel, $Mn_{1.5}Co_{1.5}O_4$, coatings have been applied onto chromia-forming metallic interconnects to prevent chromium poisoning[1]. In addition, to reduce interfacial resistance, contact paste, such as $(La_{0.9}Sr_{0.1})_{0.98}MnO_{3+\delta}$ (LSM10), is often applied between the cathode and the interconnect. Consequently, additional interfaces are created during cell fabrication. These interfaces include those between the cathode and the contact paste, between the contact paste and $Mn_{1.5}Co_{1.5}O_4$ spinel coating, between the $Mn_{1.5}Co_{1.5}O_4$ spinel coating and the thermally-grown oxide (TGO) on the surface of the metallic interconnect and between the TGO and the metallic interconnect (e.g.- AL 441, Crofer).

The risk of failure of SOFCs will increase with the number of interfaces and the potential for debonding of these interfaces has raised concerns about its impact on the long-term performance of SOFCs. Interfacial debonding on the cathode region of SOFCs could be induced by stresses that arise from mismatches in thermoelastic properties, from phase transformations and/or from the formation of new phases as a result of chemical reactions including oxidation. Knowledge about the magnitude and origin of these stresses is essential to develop models to predict the service life of SOFCs and to guide the innovation of more durable and reliable SOFCs. In this paper, we present results from experiments designed to quantify residual stresses associated with the deposition of spinel protective coating on alloys AL441 and Crofer along with results of mechanical evaluation of interfaces between LSM-10 and $Mn_{1.5}Co_{1.5}O_4$ spinel coatings.

EXPERIMENTAL PROCEDURES

Two type of interconnect materials, Crofer 22 APU (ThyssenKrupp VDM USA, Inc, Glorham, NJ) and AL441 stainless steel (Allegheny Ludlum Corporation, Brackenridge, PA), were investigated. The chemical compositions of both alloys are listed in table 1.

Table 1. Chemical composition (wt.%) of the stainless steel substrates

	Cr	Fe	C	Mn	Si	Cu	Al	S	P	Ni	Ti	Nb	La
AL441	18	Bal.	0.01	0.35	0.34	---	0.05	0.002	0.023	0.30	0.22	0.5	---
Crofer 22 APU	20.0~ 24.0	Bal.	0.03	0.3~0.8	0.5	0.5	0.5	0.020	0.050	---	0.03 ~0.2	---	0.04 ~0.2

The protective coating with a nominal composition of $Mn_{1.5}Co_{1.5}O_4$ was synthesized at Oak Ridge National Laboratory by combustion synthesis. The metal substrates were coated with this $Mn_{1.5}Co_{1.5}O_4$ using screen-printing techniques. The average thickness of the coatings was 15 μm. The sintering process of the protective coating follows procedures developed elsewhere[1]. Prior to sintering the samples were baked at 360°C for 30 minutes in air to burn out the binder. The samples were then reduced at 850°C for 4 hours under a constant flow of wet 4%H_2 96%N_2 gas. After the reduction process, the coating consists of MnO and Co [1,2]. The samples were then heated up in air to 800°C, and the desired cubic spinel phase formed. Subsequently, the residual stresses in the coating were evaluated in air at 800°C.

On top of the $Mn_{1.5}Co_{1.5}O_4$ spinel coating, $(La_{0.9}Sr_{0.1})_{0.98}MnO_{3+\delta}$ (LSM10) (Fuel Cell Materials Lewis Center, Ohio) contact paste was applied and synthesized. To evaluate the interfacial toughness, a modified four-point bend single edge notched beam (SENB) method[3-5] was successfully employed[2]. In this work, we have used this approach to evaluate symmetric sandwich test specimen geometry. Details on sample preparation are presented in the following subsections.

Residual Stress Measurement

In-situ X-ray diffraction (XRD) technique (the $sin^2\psi$ method[6]) was used to determine the residual stresses in the spinel coating on AL441 and Crofer substrates at elevated temperature. The samples were 15mm×15mm coupons with thickness of ~0.5mm. One spinel coated AL441 and one spinel coated Crofer were selected to perform the in-situ residual stress measurement. The measurements were conducted at the synchrotron x-ray beamline X14A, which is operated by the High Temperature Materials Laboratory (HTML) at the National Synchrotron Light Source. The high intensity of the monochromatic x-ray beam (~10^{12} photons/s) allows the detection of minor phases at shorter time. A Ge (111) analyzer was used to measure the stress in the parallel optics mode so that any sample surface displacement effects on peak position were eliminated[7]. The beam size at the surface of the test specimen was 0.75mm×2mm. The X-Ray energy was set to 7 keV (1.7712 Angstroms) to avoid the iron edge absorption.

Residual stresses were determined with multiple ψ (i.e., the angle between the normal to the surface of the sample and the normal to the diffraction plane) tilts[6]. In the $sin^2\psi$ method, it was assumed that the $Mn_{1.5}Co_{1.5}O_4$ coating layer was under an equibiaxial plane stress condition and that the stress state is uniform throughout the layer. According to these assumptions, the inter-planar spacing, d, has a linear relationship with $sin^2(\psi)$, where ψ is the tilt angle, and the residual stress is proportional to the slope of the d vs. $sin^2(\psi)$ plots. We also assume that the time interval for each stress measurement was short enough so that there was no significant shift in Bragg peak positions due to phase transformations or changes in chemical composition during the scan.

Interfacial Toughness Measurement

The modified four-point bend SENB technique was used for interfacial toughness evaluation. The inner and outer spans of the test fixture were 10 mm and 20 mm respectively. All the tests were performed at room temperature under a constant displacement rate of 0.5 μm/s. Test specimens were prepared according to the following procedure. First, metallic strips with a nominal thickness of 300-μm were machined to obtain the desired thickness and consistent surface roughness (Ra ~0.3 micron) and flatness (the radii of curvature were ~0.9m for 15-mm long strips and ~0.5m for 30-mm long strips). Then, $Mn_{1.5}Co_{1.5}O_4$ spinel coatings were applied onto the strips of Crofer or AL441 following the reduction/oxidation process explained above. Finally, one 30-mm and two 15-mm long coated strips were sandwiched with a layer of LSM10 contact paste in between (see the schematics in Figure 1). An artificial notch was created between the two 15-mm long metallic strips. No pore former was used in the LSM10 contact paste. The paste was prepared by mixing LSM10 powder with 15 wt% ESL 450 binder in a three roll mill. During sample preparation it is inevitable that the contact paste will overflow and fill up the notch. However, this does not affect the function of the notch for crack initiation.

McCarthy, et al., showed[8] that the pO_2 cycling process can enhance densification over regular sintering, so two conditions for sintering LSM10 were investigated: sintering in air at 900°C for 4 hours or sintering according to a cyclic PO_2 process. It was expected that the cyclic pO_2 process would improve the mechanical properties of the material. In the cyclic PO_2 process[8], the oxygen partial pressure was cycled between ~10^{-6} (pure N_2) and 0.21 (air). The duration of each cycle was 20 minutes with 10 minutes of exposure to each gas.

The nominal dimensions of the test specimens were 30mm x 3mm x 0.7mm. An inert weight, exerting a pressure of ~283Pa, was placed on top of the test specimens to ensure they remained flat during the sintering process. The geometry of the four-point bend SENB test specimen is schematically shown in Figure 1.

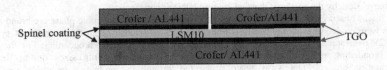

Figure 1. Schematic of the four-point bend SENB test specimen

The thickness of the $Mn_{1.5}Co_{1.5}O_4$ protective layer was 15±5 μm while the LSM10 contact paste layer varied between 60 μm and 70 μm. The interfacial toughness was calculated according to[3,4]:

$$G = \frac{(1-v^2)M^2}{2E}\left(\frac{1}{I_{c2}} - \frac{1}{I_{c5}}\right)$$

and, $M = Pl/2b$;

where, P is the load for steady crack propagation, l is half of the difference between the outer span and inter span, b is the sample width, E is Young's modulus of the metallic strip, and v is Poisson's ratio. I_{c5} is the moment of inertia of the original composite beam and I_{c2} is the moment of inertia of the remaining part beneath the crack. These equations were initially developed for a three-layer system[3,4]. We have extended the equations to five layer system and the corresponding equations are:

$$I_{c5} = \frac{1}{12}(2h^3 + 2\frac{E_1}{E}h_1^3 + \frac{E_2}{E}h_2^3) + \frac{h}{2}(2h_1 + h_2 + h)^2 + \frac{h_1 E_1}{2E}(h_1 + h_2)^2$$

$$I_{c2} = \frac{1}{12}(h^3 + \frac{E_1}{E}h_1^3) + h(\frac{h^2 + \frac{E_1}{E}h_1^2 + 2\frac{E_1}{E}hh_1}{2h + 2\frac{E_1}{E}h_1} - \frac{h}{2})^2 + \frac{h_1 E_1}{E}(\frac{h_1}{2} + h - \frac{h^2 + \frac{E_1}{E}h_1^2 + 2\frac{E_1}{E}hh_1}{2h + 2\frac{E_1}{E}h_1})^2$$

where, h_1 and h_2 refer to the layer thickness and E_1 and E_2 refer to the elastic modulus of spinel coating and the LSM10 layer. The above solution is in agreement with a recent analysis of this configuration[9]. Because the TGO layer is considerably thin $(1\sim2\mu m)$[1], it has been ignored in these calculations. At room temperature, the residual stresses in each layer are also assumed to be zero in order to calculate the interfacial toughness using the above equations; however, this assumption may not reflect the conditions of the layers.

RESULTS AND DISCUSSIONS

Residual Stress in the Oxide Spinel Layer
After the reduction process, the protective coating was transformed into MnO and metallic Co. Upon heating the sample to 800°C in air, the coating transformed into a single phase of cubic spinel, i.e., $MnCo_2O_4$, and is consistent with previously reported results[1,2]. In Figure 2, the XRD patterns of spinel coated AL441 and Crofer at room temperature and at 800°C show that the coatings have the same phases present regardless of the substrate. During cooling, the tetragonal spinel phase formed in the coating is also independent of the substrate[2]. This phase transformation was confirmed to be a reversible process by the previous thermal cycling XRD experiments[2].

It was observed that the peaks for both tetragonal phase and the cubic phase are broad after the samples were cooled to room temperature. For some peak positions (indentified with the ★ symbol in Figure 2), overlapping peaks of the two phases may cause additional broadening. For the distinct peaks, broadening may be due to the formation of small crystallite sizes during the phase transformation process. The X-ray beam penetrated through the protective layer and reached the metallic substrate, and the peaks associated with the metallic substrate are indicated with arrows in Figure 2. These results indicate that the presence of the thermally-grown oxide (TGO) layer that formed between the spinel protective layer and the metallic substrate could not be detected.

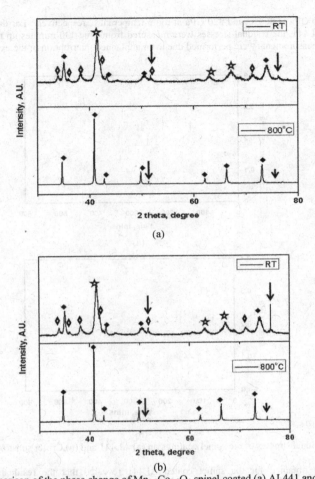

(a)

(b)

Figure 2. Comparison of the phase change of $Mn_{1.5}Co_{1.5}O_4$ spinel coated (a) AL441 and (b) Crofer at room temperature and 800°C . ◆: cubic $MnCo_2O_4$ phase; ◗: tetragonal Mn_2CoO_4 phase; ★ overlapping peaks of the $MnCo_2O_4$ and Mn_2CoO_4 phases; the peaks associated with the metallic substrate metal are identified with arrows.

The residual stresses in the coating layer were measured in air at 800°C through seven ψ angles. After scanning of the complete spectrum for 2theta range of 30 to 110 within the first 130 minutes, residual stress measurements were performed for each sample. Each stress measurement lasted about 40-50 minutes, and the residual stress results are presented in Figure 3 along with the corresponding time stamp. For illustration purposes we have assumed the elastic modulus of the spinel coating was 200 GPa at 800°C and 320 GPa at room temperature, respectively. For the $Mn_{1.5}Co_{1.5}O_4$ spinel-coated AL441, the residual stresses were measured from time 130 minutes up to 300 minutes, and no further measurements were performed due to an unplanned interruption of the experiment.

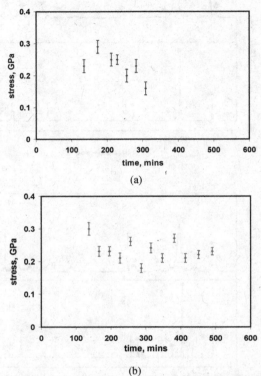

(a)

(b)

Figure 3. Residual stresses of the spinel coatings on (a) AL441 and (b) Crofer substrates at 800°C

The data obtained for the spinel coated AL441 revealed that the residual stress in the $Mn_{1.5}Co_{1.5}O_4$ spinel layer was tensile at 800°C and that it appeared to decrease with time. The last data point taken corresponds to a residual stress level of 0.16±0.02 GPa. The combination of the sintering process of the coating, the possible development of a TGO layer between the coating and the substrate and the creep deformation of the metallic substrate would result in a complex state of stress. The tensile nature of the residual stress may be associated with the shrinkage of the coating layer as a result of sintering. For the $Mn_{1.5}Co_{1.5}O_4$ spinel coated Crofer sample, the residual stresses initially decreased until it reached a constant value of 0.23±0.01 GPa for the remainder of the testing period. While differences in the creep behavior of the metallic substrate could explain the different stress relaxation behavior for both systems, the exact source of this difference in behavior can not be explained at this time. After the test specimens were cooled down to room temperature, accurate residual stresses measurements could not be performed due to the large error resulting from extensive peak broadening. However, the residual stress in the $Mn_{1.5}Co_{1.5}O_4$ spinel layer appears to be compressive for the Crofer sample. The thermal expansion coefficients (CTE) of the metal substrate and the $Mn_{1.5}Co_{1.5}O_4$ spinel are 12.6 and 11.4×10^{-6} K^{-1}, respectively, from room temperature to 800°C[1]. This CTE mismatch could be big enough to result in a compressive residual stresses in the spinel coating at room temperature.

For the operation of SOFCs, small or no residual stresses are preferred for long term operation. When the tensile stress level for the spinel coating is high enough to reach/exceed its biaxial strength, it may cause cracking of the coating, providing channels for Chromium migration. Therefore, understanding the development of the residual stresses and quantifying the stress levels are necessary to ensure the reliable operation of SOFCs.

Interfacial Toughness Measurement

A total of eight Crofer-based samples were tested for each of the two sintering conditions. An example of the load-displacement curve recorded during a four-point bend test is shown in Figure 4. In the calculations using the equations presented above, we have assumed value for the elastic modulus of 200 GPa, 320 GPa and 20 GPa for the metal substrate, spinel coating and the LSM10, respectively. For test specimens sintered in air at 900°C for four hours, the steady state load was found to be 1.60 ± 0.08 N for the samples sintered in air and 1.70 ± 0.10 N for the samples sintered under the cyclic pO_2 treatment. And consequently, the strain energy release rate was found to be 1.52 ± 0.11 J/m^2 for the sample sintered in air, whereas it had a value of 1.47 ± 0.15 J/m^2 for test specimens sintered with cyclic pO_2 treatment. A "t-test" revealed that at a 95% confidence level there are no significant differences in the interfacial toughness for these two sintering condition.

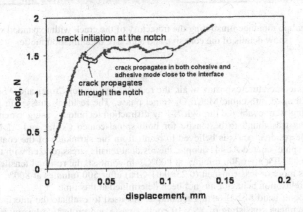

Figure 4. An example of the load-displacement curve for the 4 point bend SENB sample

Observations on the polished edge of test specimens after testing revealed details of the crack propagation path. Although the notch was filled with contact paste, it was fully functional as a notch. The reason is that the LSM is a much weaker material than the metal strips and cracks always originated inside the notch region during bending tests. As shown in Figure 5, after the crack initiated at the notch, it eventually reached the interface between LSM10 and the spinel layer. The crack propagation resulted in adhesive failure between LSM10 and the spinel layer and in cohesive failure within the LSM10 layer close to the interface. This observation indicated that improvement of the mechanical strength of the LSM10 contact paste could lead to improve the interfacial toughness. In addition, there was no damage to the spinel coating or the interface between the spinel coating and the metal substrate after testing.

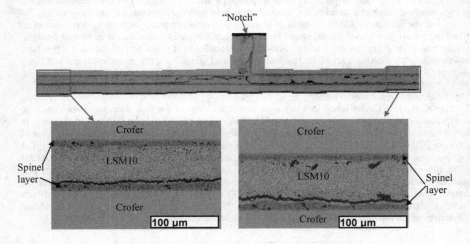

Figure 5. Micrograph montage illustrating the trajectory of the crack with expanded views of ends to show details of the cohesive and adhesive delamination mode.

SUMMARY

Upon high temperature exposure to air, the reduced coating for Crofer and AL441 interconnect materials is transformed into cubic $MnCo_2O_4$ spinel phase. The residual stresses in this cubic spinel phase of the coating were evaluated through X-Ray diffraction techniques using synchrotron radiation. The residual stress was found to be tensile for both spinel-coated Crofer and AL441 substrates at 800°C. The tensile residual stress is believed to result from the shrinkage of the coating layer during sintering. For the spinel-coated AL441sample, the residual tensile stress relaxed with time and reached a value of 0.16±0.02 GPa after 300 minutes at 800°C. In contrast, the residual tensile stress in spinel-coated Crofer was more or less constant (0.23±0.01 GPa) after 500 minutes at 800°C. The reason for this difference in relaxation behavior has not been determined at this time.

The four-point bend SENB testing technique was used to evaluate the interfacial toughness of sandwich test specimens consisting of LSM10 contact paste and protective layers, sintered at 900°C for four hours. For the two sintering conditions investigated, the strain energy release rate was found to be 1.52±0.11 J/m^2 for regular sintering and a value of 1.47 ± 0.15 J/m^2 for sintering with cyclic PO_2 treatment. It was determined that this difference was not significant.

ACKNOWLEDGEMENTS

This research was sponsored by the US Department of Energy, Office of Fossil Energy, SECA Core Technology Program at ORNL under Contract DEAC05- 00OR22725 with UT-Battelle, LLC. Work at the X14A beamline at the National Synchrotron Light Source was sponsored by the U. S. Department of Energy, Office of Energy Efficiency and Renewable Energy, Vehicle Technologies Program through the Oak Ridge National Laboratory's High Temperature Materials Laboratory User Program. The authors are indebted to John J. Henry, Claire Chisholm, Randy J. Parten and Tim R. Armstrong for their help with sample preparation and to Roberta A. Meisner and Dr. Chun-Hway Hsueh for reviewing this paper and providing valuable discussions. We are grateful to Amit Shyam for his help on interfacial testing. We greatly appreciate James Rakowski (ATI Allegheny Ludlum) for providing AL441 materials, and Larry R. Pederson, Benjamin P. McCarthy and Elizabeth Stephens (Pacific Northwest National Laboratory) for their generous help with sample preparation.

REFERENCES

[1] Z. Yang, G. Xia, and J. W. Stevenson, Electrochemical and Solid-State Letters, 8 (3) A168 (2005)
[2] SECA Core Technology Program. ORNL report for the 3rd quarter of FY08
[3] I. Hofinger, M. Oechsner, H. Bahr and M. V. Swain, Inter. J. Fract., 92, 213(1998)
[4] N. W. Klingbeil and J. L. Beuth, Eng. Fract. Mech. 56[1], 113(1997)
[5] G. Delette, J. Laurencin, M. Dupeux and J.B. Doyer, Scripta Materialia 59(1), 31 (2008)
[6] I.C. Noyan and J.B. Cohen, Residual Stress Measurement by Diffraction and Interpretation (Springer-Verlag, New York)
[7] T. R. Watkins, O. B. Cavin, J. Bai and J. A. Chediak, "Residual Stress Determinations Using Parallel Beam Optics," pp. 119-129 in Advances in X-Ray Analysis, V. 46 CD ROM. Edited by T. C. Huang et al., ICDD, Newtown Square, PA, 2003.
[8] B. P. McCarthy, L. R. Pederson, H. U. Anderson, et al., J. Am. Ceram. Soc., 90 (10), 3255(2007)
[9] C.H. Hsueh, W.H. Tuan, W.C.J. Wei, Scripta Materialia, in press.

Materials Synthesis

SYNTHESIS AND CHARACTERIZATION OF OXIDE NANOPARTICLES FOR ENERGY APPLICATIONS

Joysurya Basu, , Jonathan P Winterstein, Sanjit Bhowmick and C. Barry Carter
Department of Chemical, Materials and Bio-molecular Engineering (CMBE) and Connecticut
Global Fuel Cell Center (CGFCC), University of Connecticut, Storrs, CT, 06269

ABSTRACT

Oxide nanoparticles of compounds containing cerium ions have a number of applications for
catalysis and energy. In the present work crystalline pure and doped cerium oxide (CeO_2) and
barium cerate ($BaCeO_3$) have been synthesized using a low-temperature aqueous-solution
method. The cerium oxide has been doped with gadolinium and lanthanum and the barium cerate
has been doped with zirconium and cobalt. Transmission electron microscopy has been
performed to determine the structural and chemical nature of the nanoparticles. Observations of
surfaces faceting parallel to {001} and {111} are discussed.

INTRODUCTION

Energy-generation technologies that do not rely on fossil-fuel combustion are becoming
important for a variety of reasons including geopolitical tensions associated with the supply of
resources, emissions of nitrogen- and sulfur-containing pollutants during combustion and
production of the greenhouse gas carbon dioxide. The use of hydrogen as an energy carrier
presents a possible means by which fossil-fuel consumption can be reduced or eliminated.
Cleanly produced hydrogen can be used to power fuel cells and directly produce electricity
(avoiding combustion) with water as the only byproduct.

Many options exist for hydrogen production. Hydrogen can be produced from water using a
thermochemical redox reaction with metal oxides. For a completely clean process, the required
heat can be supplied by solar reactors [1, 2]. Solar thermal hydrogen production using metal
oxides has been demonstrated using a variety of materials including ZnO [1, 3], ferrite spinels [1,
4], and cerium oxide [5, 6]. Pilot plants for solar thermal hydrogen production have been built in
several countries. Among other benefits, solar thermal hydrogen can be produced efficiently and
only requires water and solar heat as inputs if the metal oxides are recycled.

Another clean option for hydrogen production consists of the reforming of biomass-derived
alcohols followed by separation and purification of the hydrogen. The inherent problem with this
process is the simultaneous generation of several byproducts. So this process requires the
additional step of hydrogen separation. The presence of carbon monoxide and carbon dioxide in
the hydrogen fuel for polymer-electrolyte membrane fuel cells degrades the fuel cell
performance [7]. Palladium and its alloys have been investigated as a potential material for
hydrogen separation and purification, but the materials cost and corrosion limit this technology
[7]. Hydrogen generation and separation can be achieved simultaneously by multifunctional
ceramics, which act as catalysts for the reforming step and separation membranes due to protonic
conductivity [8].

For catalysis and hydrogen generation, ceramic materials are often the key component. CeO_2 and
Ce-containing compounds are attractive materials for these applications. $BaCeO_3$ and Ba-doped

CeO_2 have already been reported as good catalysts for biofuel reaction. The structure and chemistry of these surfaces of these materials is critical. Cerium oxide is used in other energy-related applications such as automobile catalytic converters [9], catalyst supports [10] and as the electrolyte in solid oxide fuel cells [11, 12]. Au nanoparticles supported on CeO_2 is an active catalyst system for converting CO to CO_2 [13, 14]. Cerium oxide is also an important chemical-mechanical polishing medium [15]. Nanostructured ceramics based on $BaCeO_3$ and doped-CeO_2 have the additional advantage of a high surface-to-volume ratio. Efficiency of catalytic converters can be enhanced by depositing Au nanoparticles on CeO_2-based nanostructures as the number density of the catalyst particles will be greatly increased. For production of dense ceramic membranes, the use of nanoparticles as the starting material can greatly reduce sintering times and temperatures. In some cases, low-temperature sintering may be necessary to avoid formation of unwanted high-temperature phases.

Various synthetic routes to prepare crystalline pure, doped-CeO_2 and $BaCeO_3$ have been explored. All these routes require high temperature and pressure. A low-temperature route to prepare these materials would be more cost-effective. Additional advantages sought in a synthesis route are ease of doping and avoidance of a subsequent crystallization step. In the present paper, a simple low-temperature synthesis route to prepare pure, doped-CeO_2 and $BaCeO_3$ as crystalline nanoparticles is reported. Applicability of these materials along with the advantages over the other synthetic routes will be discussed.

EXPERIMENTAL DETAILS

Pure and doped CeO_2 and $BaCeO_3$ have been synthesized by a hydrothermal reaction of analytical grade nitrate salts of the metals with hexamethylene tetramine at 80 °C. For the synthesis of CeO_2, a 0.025 M aqueous solution of cerium nitrate hexahydrate ($Ce(NO_3)_3$, $6H_2O$, Alfa Aesar) was mixed with 0.025 M aqueous solution of hexamethylene tetramine (HMTA) ($C_6H_{12}N_4$, Alfa Aesar) at 60 °C and the solution was kept at 80 °C for about 45 minutes. The solution turns milky in about 6-9 minutes at 80 °C, which is considered to be the start of the reaction. For doping CeO_2 with La and Gd by 10 at% separately, 0.0025 M aqueous solution of Lanthanum nitrate hexahydrate ($La(NO_3)_3$, $6H_2O$, Alfa Aesar) and 0.0025 M aqueous solution of gadolinium nitrate hydrate ($Gd(NO_3)_3$, xH_2O, Alfa Aesar) were mixed with the cerium nitrate solution and mixed solution was reacted with HMTA at 80 °C for about 45 minutes.

For the synthesis of pure $BaCeO_3$ 0.025 M aqueous solution of barium nitrate (Alfa Aesar) was mixed with equal volume of 0.025 M aqueous solution of cerium nitrate hexahydrate and 0.025 M aqueous solution of HMTA at 60 °C and the solution was heated at 80 °C for 45 minutes. For doping $BaCeO_3$ with Zr and Co cerium nitrate concentration was reduced by 10 at% and it was replaced with 0.0025 M aqueous solution of zirconium nitrate oxyhydrate (Sterm Chemicals). Similarly for Co doping the concentration of cerium nitrate was reduced by 10 at% and it was replaced with 0.0025 M aqueous solution of cobalt nitrate (Johnson Matthey Chemicals Limited). The addition of dopants slows down the reaction kinetics considerably.

The particles so produced were characterized by a JEOL 2010 FasTEM and a Tecnai T12 TEM. For TEM observation the particles were washed twice and about three drops of the solution was taken onto a ultrathin carbon grid.

RESULTS AND DISCUSSION

A high-angle annular detector dark-field (HAADF) STEM image of the pure CeO_2 and the bright-field STEM image of La-doped CeO_2 particles are given in figure 1(a-b) respectively. HAADF is used to understand the particles shape and morphology clearly. It is observed from the micrographs that the pure CeO_2 particles are cuboidal in shape and they are in the size range of 10-25 nm. The particles are crystalline as can be confirmed from the electron diffraction pattern. The particles are often clustered. Doping with La does not change the crystallinity, morphology and the size of the particles.

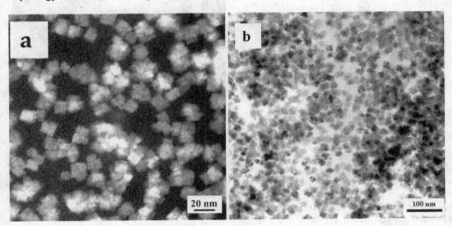

Figure 1: (a) HAADF image of pure CeO_2 (b) STEM image of La-doped CeO_2 nanoparticles. The particles are cuboidal and 10-25 nm in size.

Doping the particles with Gd resulted in a reduced size (by about 5-10 nm). High-resolution TEM observation of the particles reveals that the doped particle surfaces are not smooth. Atomically flat {111} and {100} facets can be seen on the particles. Time-resolved high-resolution TEM indicates that there is a dynamic equilibrium between {111} and {100} facets, which can be attributed to the vacancy and ionic diffusion along the particle facets. Addition of Gd does change the particle morphology and most of the Gd-doped particles show {111} and {100} facets.

A TEM bright-field image and the electron diffraction pattern of the $BaCeO_3$ nanoparticles are given in figure 2(a-b). It is observed from the image that the particles are again cuboidal and 10-25 nm in size. Presence of the Debye rings in the diffraction pattern confirms that the particles are crystalline. Zr-dpoing reduces the particle size by about 5-10 nm and the particle morphology is also changed. Co-doping does not change the particle morphology as such but the size is reduced by 5-10 nm.

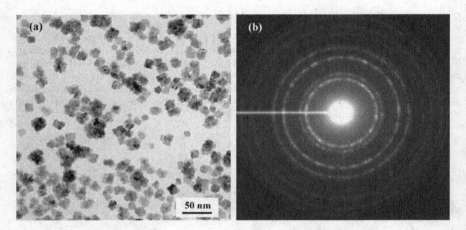

Figure 2: TEM BF image (a) and SAD (b) of the hydrothermally synthesized BaCeO₃ nanoparticles. The particles are cuboidal in shape and 10-20 nm in size.

X-ray and electron diffraction signatures of pure CeO_2 and $BaCeO_3$ are quite similar. In fact it is often difficult to distinguish them from the diffraction pattern. For energy and catalytic applications chemistry of the nanoparticles is quite important along with the structure. In fact Ba-doped CeO_2 is also known to be catalytically very active material. It is pointed out in this paper that the authors have been able to synthesize cubic nanoparticles of similar chemistry via low-temperature hydrothermal route.

High-resolution TEM image of two different $BaCeO_3$ particles are given in figure 3(a-b). Lattice planes are clearly resolved in the images. It is observed that the particles are often faceted along {111} planes and that they often sinter to other particles by sharing a common facet and thus they create a $\Sigma=1$ grain boundary.

Doping the $BaCeO_3$ nanoparticles with Zr and Co in proper stoichiometry in order to achieve the catalytic activity and protonic conductivity, results in a reduced size (by about 5-10 nm). After Zr-doping the particle morphology is changed and a different faceting behavior can be observed. Co-doping does not change the particle morphology considerably. It is understood from this observation that surface energy plays an important role in the nucleation and growth of the nanoparticles from the aqueous solution. The study of the catalytic activity and the conduction mechanism is in progress.

The growth behavior of the pure- and doped-CeO_2 and $BaCeO_3$ in the aqueous solution has been studied by continuing the reaction for longer time. In both the cases the particles do not grow beyond 20-30 nm. It is postulated that a large number of nuclei forms at the start of the nucleation process and as a very dilute solution has been used for processing, unavailability of metal ions restricts the growth of the nanoparticles. The reaction kinetics is strongly dependent on the chemistry of the solution. The nucleation of pure CeO_2 starts in about 6-9 minutes at 80

^{0}C. Addition of dopants slows down the kinetics and often the reaction starts after keeping the reacting solution for 15-18 minutes at 80 ^{0}C. High-resolution electron microscopy and image simulation indicate that the particles grow by ionic attachments on the surfaces.

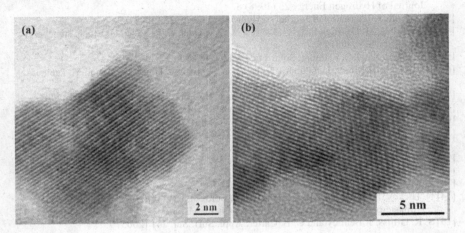

Figure 3(a-b): HRTEM image of the BaCeO$_3$ particles. The particles are crystalline and faceted.

CONCLUSIONS

The authors have successfully synthesized nanocrystalline cuboidal CeO$_2$ and BaCeO$_3$ at low temperature. The particles are formed in the crystalline state. Furthermore, the particles can be doped successfully with La, Gd, Zr and Co. Often the particle size is reduced and the shape of the particle is changed by doping. This is one of the early works on the low-temperature hydrothermal synthesis of pure and doped CeO$_2$ and BaCeO$_3$. It is further understood that the particles grow by ionic attachments on the surfaces and surface energies play an important role in determining the particle shapes. It is possible to grow particles of different shapes by changing the surface energies.

FUTURE WORK

It has been planned to consolidate the particles in the form of a porous membrane so that the membrane can act both as a catalyst for reforming process and a protonic conductor for hydrogen separation. It has been also planned to deposit Au nanoparticles on the surfaces to obtain a higher number density of Au and use it as a catalytic support for CO to CO$_2$ conversion.

ACKNOWLEDGEMENTS

The authors thank Dr Roger Ristau and Dr. Virgil Solomon and both Aravind Suresh and Prof. Ben Wilhite for helpful discussions. Support from the University of Connecticut is thankfully acknowledged.

REFERENCES

[1] H. Kaneko, N. Gokon and N. Hasegawa, Energy 30 (2005) 8.
[2] A. Steinfeld, P. Kuhn, A. Reller, R. Palumbo, J. Murray and Y. Tamaura, International Journal of Hydrogen Energy 23 (1998) 8.
[3] A. Steinfeld, International Journal of Hydrogen Energy 27 (2002) 9.
[4] J. E. Miller, M. D. Allendorf, R. B. Diver, L. R. Evans, N. P. Siegel and J. N. Stuecker, J Mater Sci 43 (2008) 15.
[5] S. Abanades and G. Flamant, Solar Energy 80 (2006) 13.
[6] H. Kaneko, T. Miura, H. Ishihara, S. Taku, T. Yokoyama, H. Nakajima and Y. Tamaura, Energy 32 (2007) 8.
[7] B. A. Wilhite, M. A. Schmidt and K. F. Jensen, Ind. Eng. Chem. Res. 43 (2004) 7083.
[8] A. Suresh, J. Basu, B. A. Wilhite and C. B. Carter, Mat. Res. Soc. Symp. Proc. (2009).
[9] H. Muraki and G. Zhang, Catalysis Today 63 (2000) 9.
[10] Z.-R. Tang, J. K. Edwards, J. K. Bartley, S. H. Taylor, A. F. Carley, A. A. Herzing, C. J. Kiely and G. J. Hutchings, J. Catal. 249 (2007) 12.
[11] B. C. H. Steele, Journal of Materials Science 36 (2001) 16.
[12] S. M. Haile, Acta Mater 51 (2003) 20.
[13] Z. Chen and Q. Gao, Appl. Catal. B 84 (2008) 7.
[14] N. Hickey, P. A. Larochette, C. Gentilini, L. Sordelli, L. Olivi, S. Polizzi, T. Montini, P. Fornasiero, L. Pasquato and M. Graziani, Chem. Mater. 19 (2007) 2.
[15] S. R. Gilliss, J. Bentley and C. B. Carter, Appl. Surf. Sci. 241 (2005) 7.

GLYCINE-NITRATE SYNTHESIS AND CHARACTERIZATION OF Ba(Zr$_{0.8-x}$Ce$_x$Y$_{0.2}$)O$_{2.9}$

R. R. Chien[a]*, V. Hugo Schmidt[a], S.-C. Lee[b], C.-C. Huang[b], and Stachus P. Tu[c]
[a]Department of Physics, Montana State University, Bozeman, MT 59717, USA
[b]Department of Physics, Fu Jen Catholic University, Taipei, Taiwan 242, R.O.C.
[c]Bozeman High School, Bozeman, MT 59715, USA

ABSTRACT
    A combustion method, the glycine-nitrate process, has been used to synthesize nanosized Ba(Zr$_{0.8-x}$Ce$_x$Y$_{0.2}$)O$_{2.9}$ (BZCY) ceramic powders. As-synthesized and calcined Ba(Zr$_{0.7}$Ce$_{0.1}$Y$_{0.2}$)O$_{2.9}$ (BZCY712) ceramic powders were investigated as functions of glycine-to-nitrate molar (G/N) ratio (G/N=0.33, 0.40, 0.45, 0.50, 0.67, 0.75, 1.00). Structures and phase transformations of calcined Ba(Zr$_{0.8-x}$Ce$_x$Y$_{0.2}$)O$_{2.9}$ (x=0.0-0.8) ceramic powders synthesized with G/N=0.5 were investigated by *in situ* x-ray diffraction (XRD) and Raman scattering. It was found that the G/N ratios near and slightly above stoichiometric ratio (5/9) yield the optimal single phase in as-synthesized BZCY712 powders. Particle sizes of as-synthesized (with G/N ratio of 0.5) Ba(Zr$_{0.8-x}$Ce$_x$Y$_{0.2}$)O$_{2.9}$ (x=0.0-0.8) powders vary from 5 to 15 nm, and increase to 30-40 nm after calcination. The calcined Ba(Ce$_{0.8}$Y$_{0.2}$)O$_{2.9}$ powder exhibits a phase transformation from possibly mixed rhombohedral/monoclinic phase to rhombohedral phase in the region of 300-350 °C, then to a cubic phase near 700 °C.

INTRODUCTION

High temperature proton conducting BaZrO$_3$-, SrZrO$_3$-, CaZrO$_3$-, BaCeO$_3$-, SrCeO$_3$-, and CaCeO$_3$-based ABO$_3$ perovskite-type oxides exhibit good protonic conduction under hydrogen-containing atmosphere at elevated temperatures.[1,2] They are promising materials for applications of proton conducting solid oxide fuel cells, hydrogen separation membranes, hydrogen pumps, hydrogen sensors, and steam electrolyzers for hydrogen production.[3] In general, cerate-based protonic conductors have a high conductivity but exhibit poor chemical stability in CO$_2$- and H$_2$O-containing atmosphere at elevated temperature.[4,5] Although zirconate-based protonic conductors containing larger grain boundary contributions have lower ionic conductivity, they show very good chemical and mechanical stability.[6] These results suggest that proton conducting solid solutions between cerate and zirconate may have both high proton conductivity and good chemical stability.[7,8] A^{II}B^{IV}O$_3$-based proton conductors are doped in B-site by lower valence elements, typically Y^{3+} or trivalent rare earth metal cations, creating oxygen vacancies. Subsequent exposure to humid atmospheres is presumed to lead to the incorporation of protons, resulting in proton conduction.[9,10]

There are various synthesis techniques for high-quality BaCeO$_3$- and BaZrO$_3$-based proton conducting ceramic powders including solid-state reaction,[11,12] co-precipitation,[8] sol-gel,[13] hydrothermal[14] and combustion.[15,16] The synthesis of ultrafine powders with controlled powder characteristics is important to make dense sintered products at a lower sintering temperature.[17] Nanosized powders are essential for making nanocrystalline ceramics. Better conductivity in nanocrystalline materials compared to microstructured samples has been reported in yttrium stabilized zirconia (YSZ) and Y$_2$O$_3$-doped ZrO$_2$ due to the higher grain boundary conductivity in nanocyrstalline YSZ compared to microcrystalline YSZ samples.[18]

The glycine-nitrate process (GNP) using glycine as a fuel and nitrate as an oxidizer, attracts great interest because it can produce powders that have greater compositional uniformity, and lower residual carbon levels and nanoparticle sizes.[15] Glycine (NH$_2$CH$_2$COOH), one of the cheapest amino acids, is used as a fuel and acts as a complexing agent for metal cations as it has a carboxylic acid group at one end and amino group at the other end. The zwitterionic character of a glycine molecule can effectively complex metal cations of varying ionic size, which increases their solubility and prevents selective precipitation as water is evaporated to maintain compositional homogeneity among the constituents. The site for the complexed cation depends on the size and charge of the cation, and the pH value. It has been recognized that alkali and alkaline-earth (Ba^{+2}) cations are most effectively complexed by the carboxylic acid group, while many transition metal cations (Zr^{+4}, Y^{+3}) and rare-earth cations (Ce^{+3}) are most effectively complexed by the amine group.[15,19] The glycine to cation (G/C) or glycine to nitrate (G/N) ratio has to be considered to have sufficient glycine in the solution to retain complete complexation of the cations as the solution is being evaporated.

In this paper, nano-sized powders of proton conducting $Ba(Zr_{0.8-x}Ce_xY_{0.2})O_{2.9}$ (BZCY) ceramics have been synthesized by the GNP, and characterized by the methods of *in situ* x-ray diffraction (XRD) and micro-Raman spectroscopy.

EXPERIMENTAL PROCEDURE

The starting materials are $Ba(NO_3)_2$ (Alfa Aesar, 99.95%), $ZrO(NO_3)_2 \cdot xH_2O$ (Alfa Aesar, 99.9%), $Ce(NO_3)_3 \cdot 6H_2O$, and $Y(NO_3)_3 \cdot 6H_2O$ (Alfa Aesar, 99.9%) metal nitrate powders. Each was dissolved in deionized water to make the metal nitrate solution. The appropriate molar ratios of the nitrates and glycine were mixed to obtain a transparent solution. The solution then was heated and stirred at ~150-200 °C to evaporate excess water. The resulting viscous liquid autoignited and produced the desired powder. All calcined powders were calcined at 1400 °C for 5 hours in laboratory air, which is a moisture-containing atmosphere.

For *in situ* x–ray diffraction measurements, a high–temperature Rigaku Model MultiFlex x–ray diffractometer with Cu $K_{\alpha 1}$ (λ=0.15406 nm) and Cu $K_{\alpha 2}$ (λ=0.15444 nm) radiations was used in an argon-containing environment. The intensity ratio between $K_{\alpha 1}$ and $K_{\alpha 2}$ is about 2:1.[20] Powders were smoothly pressed on the platinum sample holder.

A double grating Jobin Yvon Model U-1000 double monochromator with 1800 grooves/mm gratings and a nitrogen-cooled CCD as a detector were employed for the post micro–Raman scattering measurements. A Coherent Model Innova 90 argon laser with wavelength λ=514.5 nm was used as an excitation source. With a notch filter (band width=8 nm), the Raman scattering was performed in the frequency region of 150–1600 cm^{-1}.

Particle sizes of the BZCY powder samples were estimated by using full width at half maxima (FWHM) of the major XRD peak with Scherrer's formula assuming spherical particles:

$$D = \frac{0.9\lambda}{\beta \cos\theta},$$

(1)

where D is the crystallite size (nm), λ is the x-ray radiation wavelength (0.15406 nm), θ is the diffraction peak angle, and β is the full width at half maxima (FWHM) of the peak. The XRD spectra were fitted by using PeakFit software with the sum of Gaussian and Lorentzian terms to obtain β. The $K_{\alpha 1}$ and $K_{\alpha 2}$ peaks for the same reflection should have the same separation and ratio of full width at half maximum.

RESULTS AND DISCUSSION

The stoichiometric GNP combustion reaction occurs according to the following complete reaction equation assuming the gaseous products to be CO_2, N_2, and H_2O:

$$Ba(NO_3)_{2(aq)} + (0.8\text{-}x)ZrO(NO_3)_{2(aq)} + xCe(NO_3)_{3(aq)} + 0.2Y(NO_3)_{3(aq)} + \alpha\, NH_2CH_2COOH_{(aq)}$$
$$\rightarrow Ba(Zr_{0.8-x}Ce_xY_{0.2})O_{2.9\ (c)} + CO_{2\ (g)} + N_{2\ (g)} + H_2O_{\ (g)}$$

(2)

According to the principles of propellant chemistry, Jain et al.[21] introduced a simple method of calculating the stoichiometric redox reaction between a fuel and an oxidizer, namely that the ratio of the net oxidizing valence of the metal nitrate to the net reducing valence of the fuel should be unity. The assumed valences are those presented by the elements in the usual products of the combustion reaction, which are primarily N_2, CO_2, and H_2O evolved as the gaseous products. Therefore, the elements carbon and hydrogen are considered as reducing elements with the corresponding valences +4 and +1, oxygen is considered an oxidizing element with the valency -2, and nitrogen is considered as having a valency of zero. The extrapolation of this concept to the combustion synthesis of ceramic oxides means that metals like barium, cerium and yttrium (or any other metals) should also be considered as reducing elements with the valences they have in the corresponding oxides, i.e. +2, +3 and +3 respectively. Therefore, the total oxidizing valency of nitrates is $-21-5x$, whereas the total reducing valency of NH_2CH_2COOH is +9, so the stoichiometric redox reaction can be expressed as:

$$Ba(NO_3)_2 + (0.8-x)ZrO(NO_3)_2 + xCe(NO_3)_3 + 0.2Y(NO_3)_3 + [(21+5x)/9]NH_2CH_2COOH + (x/4)O_2$$
$$\rightarrow Ba(Zr_{0.8-x}Ce_xY_{0.2})O_{2.9} + [(42+10x)/9]CO_2 + [(49/15)+(7x/9)]N_2 + [(105+25x)/18]H_2O \qquad (3)$$

where $\alpha = (21+5x)/9$ in Eq. (2) is the "stoichiometric glycine-to-barium nitrate molar ratio". Therefore, the stoichiometric glycine-to-nitrate molar ratio (G/N) is $\alpha/(4.2+x) = (21+5x)/(4.2+x)9 = 5/9$, because for $x=0$ there are 4.2 times as many nitrate ions as metal ions in the nitrates.[22] Note that some papers use glycine-to-metal nitrate molar ratio as glycine-to-nitrate molar ratio (G/N).[23,24] The stoichiometric ratio implies that the oxygen content of oxidants can be completely reacted to oxidize/consume the organic fuel exactly according to the propellant chemistry rules. As a result, the combustion process can produce H_2O, CO_2, and N_2 gases without the necessity of getting oxygen from outside.

The theoretical combustion temperature can be calculated from Hess's Law for the enthalpy of combustion:

$$\Delta H^0_{reaction} = \left(\sum n\Delta H^0_f\right)_{products} - \left(\sum n\Delta H^0_f\right)_{reactants} \qquad (4)$$

$$Q = -\Delta H^0_{reaction} = \int_{298}^{T} \left(\sum nC_p\right)_{products} dT \qquad (5)$$

where $\left(\Delta H^0_f\right)_{reaction}$ is the enthalpy of combustion at 298 K, n is the number of the moles, $\left(\Delta H^0_f\right)_{products}$ and $\left(\Delta H^0_f\right)_{reactants}$ represent the enthalpies of formation of products and reactants. Q is the heat absorbed by products under adiabatic condition, C_p is the heat capacity of the products at constant pressure, and T is the theoretical adiabatic flame temperature. The enthalpies of the reactants and products are listed in Table 1.[22,23]

Table 1 Relevant thermodynamic data

Compound	ΔH^0_f (kcal/mol)	Cp (cal/mol K)
$NH_2CH_2COOH_{(c)}$	-79.71	
$CO_{2 (g)}$	-94.051	10.34 + 0.00274T
$N_{2 (g)}$	0	6.50 + 0.0010T
$H_2O_{(g)}$	-57.796	7.20 + 0.0036T
$O_{2 (g)}$	0	5.92 + 0.00367T

(c) - Crystalline, (g) - gas, T - Absolute temperature

For BZCY712 with stoichiometric glycine-to-barium nitrate molar ratio $\alpha=21.5/9$, i.e. glycine-to-nitrate molar ratio G/N=5/9, the chemical reaction can be balanced as:

$$Ba(NO_3)_2 + 0.7ZrO(NO_3)_2 \cdot H_2O + 0.1Ce(NO_3)_3 \cdot 6H_2O + 0.2Y(NO_3)_3 \cdot 6H_2O + (21.5/9)NH_2CH_2COOH +$$
$$(0.1/4)O_2 \rightarrow Ba(Zr_{0.7}Ce_{0.1}Y_{0.2})O_{2.9} + (43/9)CO_2 + [(49/15)+(0.7/9)]N_2 + [(141+11.5)/18]H_2O, \qquad (6)$$

assuming one water of hydration in zirconium nitrate. With various glycine-to-barium nitrate ratios α, the chemical reaction can be expressed as follows:

$$Ba(NO_3)_2 + 0.7ZrO(NO_3)_2 \cdot H_2O + 0.1Ce(NO_3)_3 \cdot 6H_2O + 0.2Y(NO_3)_3 \cdot 6H_2O + \alpha NH_2CH_2COOH$$
$$\rightarrow Ba(Zr_{0.7}Ce_{0.1}Y_{0.2})O_{2.9} + 2\alpha CO_2 + [(4.3+\alpha)/2]N_2 + [5(\alpha+1)/2]H_2O + (10.7-4.5\alpha)O_2 \qquad (7)$$

In this study, we investigate how G/N ratio affects the glycine-nitrate process or the theoretical adiabatic flame temperature with various desired G/N ratios 1/3, 2/5, 3/7, 9/20, 1/2, 5/9, 2/3, 3/4, 1. Since total nitrate moles for BZCY712 is 4.3, α can be obtained by multiplying the desired G/N ratio by 4.3, i.e. $[G/N]_{ratio} = (\alpha/4.3)$. As α increases by $\Delta\alpha$, the change of enthalpy in products $(\Delta H_f^0)_{products}$ is $[2\Delta\alpha(\Delta H_{f,CO_2}^0) + 0.5\Delta\alpha(\Delta H_{f,N_2}^0) + 2.5\Delta\alpha(\Delta H_{f,H_2O}^0) - 4.5\Delta\alpha(\Delta H_{f,O_2}^0)]$ and the change of enthalpy in reactants $(\Delta H_f^0)_{reactants}$ is $[\Delta\alpha(H_{f,NH_2CH_2COOH}^0)]$. Then, the total change of the enthalpy resulting from the increment of glycine $\Delta\alpha$ is $\Delta(\Delta H_f^0) = [2\Delta\alpha(-94.051)+2.5\Delta\alpha(-57.796)]-\Delta\alpha(-79.710)$ $=-252.88\Delta\alpha$ (kcal/mole). Thus, the increased amount of the heat absorbed by products under adiabatic condition is $\Delta Q = 252.88\Delta\alpha$ kcal/mole. This indicates that the theoretical adiabatic flame temperature increases as stoichiometric glycine-to-barium nitrate molar ratio α or glycine-to-nitrate molar ratio (G/N) increases.

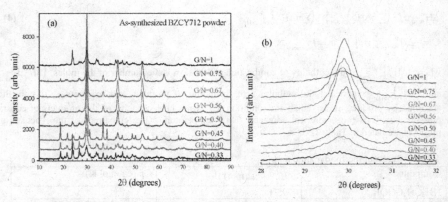

Figure 1   (a) XRD spectra of as-synthesized $Ba(Zr_{0.7}Ce_{0.1}Y_{0.2})O_{2.9}$ (BZCY712) powder synthesized with various G/N ratios. (b) Enlarged (110) peaks of the XRD spectra.

As shown in Figures 1(a)-(b), the G/N ratios near and above stoichiometric ratio 5/9≅0.56 yield the optimal single phase in the as-synthesized BZCY712 powders. As shown in Figure 2, except for the G/N=1 case the particle size of BZCY712 increases as the G/N ratio increases, i.e. the theoretical adiabatic flame temperature increases. The particle sizes of the as-synthesized BZCY712 powders synthesized with various G/ N ratio vary from 8 to 20 nm, and increase to 25-40 nm after calcinations [Figure 2]. The G/N=1 seems to produce the smallest particles, but it results from the most incomplete reaction with the greater portion of materials in black and grey as shown in Figure 3(h). Therefore, the actual flame temperature for the G/N=1 is much lower than the theoretical combustion temperature due to the large degree of incomplete reaction.

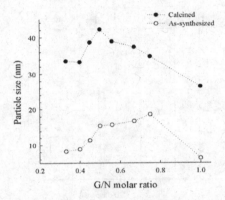

Figure 2   Particle sizes of as-synthesized and calcined Ba(Zr$_{0.7}$Ce$_{0.1}$Y$_{0.2}$)O$_{2.9}$ (BZCY712) powders as a function of G/N molar ratio.

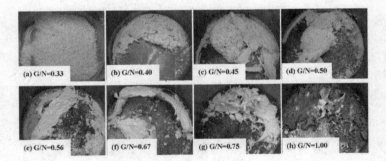

Figure 3   As-synthesized Ba(Zr$_{0.7}$Ce$_{0.1}$Y$_{0.2}$)O$_{2.9}$ (BZCY712) powders from the glycine-nitrate process with glycine-to-nitrate molar ratio (a) 1/3=0.33, (b) 2/5=0.40, (c) 9/20=0.45, (d) 1/2=0.50, (e) 5/9=0.56, (d) 2/3=0.67, (e) 3/4=0.75, and (f) 1.00.

G/N=0.5 was used for the synthesis of Ba(Zr$_{0.8-x}$Ce$_x$Y$_{0.2}$)O$_{2.9}$ (x=0.0-0.8) (BZCY) ceramic powders to investigate the GNP synthesis as a function of cerium content. The ratio G/N=0.5 yields the phase closer to the perovskite single phase for cerium molar fraction x<0.4, as shown in Figure 4(a). After calcination at 1400 °C for 5 hours in laboratory air, single phases were obtained for Ba(Zr$_{0.8-x}$Ce$_x$Y$_{0.2}$)O$_{2.9}$. [Figure 4(b)] As shown in Figure 5, particle sizes of the as-synthesized BZCY powders vary from 5 to 15 nm, and increase to 30-40 nm after the calcination.

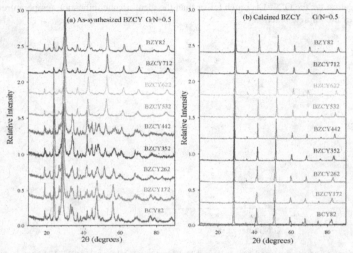

Figure 4   XRD spectra of (a) as-synthesized and (b) calcined Ba(Zr$_{0.8-x}$Ce$_x$Y$_{0.2}$)O$_{2.9}$ (x=0.0-0.8) powder synthesized with G/N molar ratio 0.5.

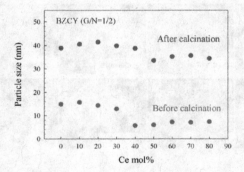

Figure 5   Particle size of as-synthesized and calcined Ba(Zr$_{0.8-x}$Ce$_x$Y$_{0.2}$)O$_{2.9}$ (x=0.0-0.8) powders as a function of Ce mole%.

As shown in Figure 6, the temperature-dependent XRD measurements in Ar-containing environment indicates that the $Ba(Zr_{0.8-x}Ce_xY_{0.2})O_{2.9}$ (x=0.6-0.8) powders seems to exhibit a phase transformation in the region of 300-400 °C. The fittings of (110) XRD peaks in Figure 7(a) for the $Ba(Ce_{0.8}Y_{0.2})O_{2.9}$ (BCY82) powders exhibiting two peaks from both $K_{\alpha 1}$ and $K_{\alpha 2}$ respectively, indicate a possibly rhombohedral (R) phase mixed with a monoclinic (M) phase, at room temperature (25 °C).[25,26] As shown in Figures 7(b) and 8(b), the d spacings calculated from (110) XRD peaks and the vibration frequency of Raman mode near 640 cm^{-1} confirm a mixed R/M to R phase transition near 300-350 °C, where M phase disappears, then a cubic phase proceeds at a higher temperature near 700 °C.[26]

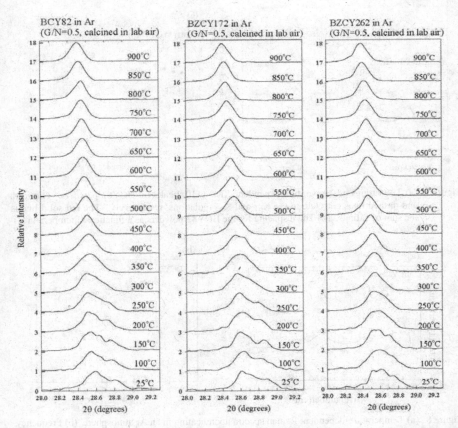

Figure 6   Temperature-dependent XRD (110) peaks of the calcined $Ba(Zr_{0.8-x}Ce_xY_{0.2})O_{2.9}$ (x=0.6-0.8) (BCY82, BZCY172, and BZCY262) powders in Ar atmosphere.

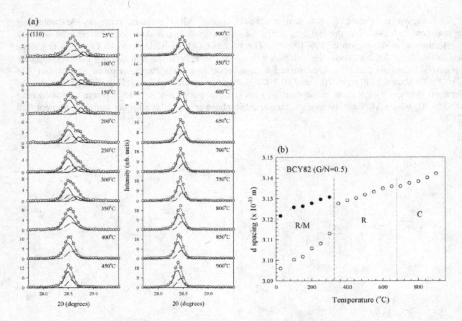

Figure 7 (a) The fittings of temperature-dependent XRD (110) peaks for BCY82 powders. The solid and dashed lines correlate to the K$_{\alpha1}$ and K$_{\alpha2}$ radiations, respectively. The red solid line is the sum of fittings. (b) The d spacing for the BCY82 powders as a function of temperature.

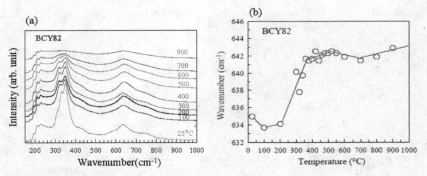

Figure 8 (a) Temperature-dependent Raman spectra upon heating in an Ar atmosphere, (b) Frequency vs. temperature of Raman mode near 640 cm^{-1}.

CONCLUSIONS

Nanosized $Ba(Zr_{0.8-x}Ce_xY_{0.2})O_{2.9}$ (x=0.0-0.8) (BZCY) ceramic powders of 5-15 nm were synthesized by the glycine-nitrate process. Particle size was investigated as a function of glycine-to-nitrate ratio (G/N) and the calcination. The theoretical adiabatic flame temperature is increased by increasing the G/N ratio. Higher flame temperature yields larger particle size of the ceramic powders. Particle size is increased to 30-40 nm after the BZCY powders were calcined at 1400 °C for 5 hours in laboratory air. Temperature-dependent measurements of XRD and micro-Raman indicate that the calcined $Ba(Zr_{0.8-x}Ce_xY_{0.2})O_{2.9}$ (x=0.6-0.8) powders exhibit a phase transformation in the region of 300-400 °C. The XRD also indicates that the calcined $Ba(Ce_{0.8}Y_{0.2})O_{2.9}$ powder exhibits a possibly mixed rhombohedral/monoclinic phase to a rhombohedral phase in the region of 300-350 °C, then to a cubic phase near 700 °C.

ACKNOWLEDGEMENT

This work was supported by DOE under subcontract DE-AC06-76RL01839 from Battelle Memorial Institute and PNNL, and National Science Council of Taiwan Grant No. 96-2112-M-030-001.

REFERENCES

[1] H. Iwahara, T. Esaka, H. Uchida, and N. Maeda, Proton Conduction in Sintered Oxides and its Application to Steam Electrolysis for Hydrogen Production, *Solid State Ionics*, **3/4**, 359-363 (1981).
[2] H. Iwahara, H. Uchida, K. Kondo, and K. Ogaki, Proton Conduction in Sintered Oxides Based on $BaCeO_3$, *J. Electrochem. Soc.*, **135**, 529-533 (1988).
[3] H. Iwahara, Technological Challenges in the Application of Proton Conducting Ceramics, *Solid State Ionics*, **77**, 289-298 (1995).
[4] N. Bonanos, K. S. Knight, and B. Ellis, Perovskite Solid Electrolytes: Structure, Transport Property, and Fuel Cell Applications, *Solid State Ionics*, **79**, 161-170 (1995).
[5] C. W. Tanner and A. V. Virkar, Instability of $BaCeO_3$ in H_2O-Containing Atmospheres, *J. Electrochem. Soc.*, **143**, (1996).
[6] K.D. Kreuer, On the Development of Proton Conducting Materials for Technological Applications, *Solid State Ionics*, **97**, 1-15 (1997).
[7] C. Zuo, S. Zha, M. Liu, M. Hatano, and M. Uchiyama, $Ba(Zr_{0.1}Ce_{0.7}Y_{0.2})O_{3-\delta}$ as an Electrolyte for Low-Temperature Solid-Oxide Fuel Cells, *Adv. Mater.*, **18**, 3318–3320 (2006).
[8] Z. Zhong, Stability and Conductivity Study of the $BaCe_{0.9-x}Zr_xY_{0.1}O_{2.95}$ Systems, *Solid State Ionics*, **178**, 213-220 (2007).
[9] T. Norby, M. Widerøe, R. Glöckner, and Y. Larring, Hydrogen in Oxides, *Dalton Trans.*, 3012-3018 (2004).
[10] K.D. Kreuer, Proton-Conducting Oxides, *Annu. Rev. Mater. Res.*, **33**, 333–359 (2003).
[11] K. Katahira, Y. Kohchi, T. Shimura, H. Iwahara, Protonic Conduction in Zr-Substituted $BaCeO_3$, *Solid State Ionics*, **138**, 91-98 (2000).
[12] K. H. Ryu and S. M. Haile, Chemical Stability and Proton Conductivity of Doped BaCeO –BaZrO Solid Solutions, *Solid State Ionics*, **125**, 355–367 (1999).
[13] R. B. Cervera, Y. Oyama, and S.Yamaguchi, Low Temperature Synthesis of Nanocrystalline Proton Conducting $BaZr_{0.8}Y_{0.2}O_{3-\delta}$ by Sol–Gel Method, *Solid State Ionics*, **178**, 569–574 (2007).
[14] W. Zheng, C. Liu, Y Yue, and W. Peng, Hydrothermal Synthesis and Characterization of $BaZr_{1-x}M_xO_{3-\alpha}$ (M=Al, Ga, In, x≤0.20) Series Oxide, *Mater. Lett.*, **30**, 93-97 (1997).
[15] L. A. Chick, L. R. Pederson, G. D. Maupin, J. L. Bates, L. E. Thomas, and G. J. Exarhos, Glycine-Nitrate Combustion Synthesis of Oxide Ceramics Powders, *Mater. Lett.*, **10**, 6-12 (1990).

[16] P. Babilo, T. Uda, and S. M. Haile, Processing of Yttrium-Doped Barium Zirconate for High Proton Conductivity, *J. Mater. Res.*, **22**, 1322-1330 (2007).

[17] Y. Zhou and M. N. Rahaman, Effect of Redox Reaction on the Sintering Behavior of Cerium Oxide, *Acta Mater.*, **45**, 3635-3639 (1997).

[18] N. H. Perry, S. Kim, and T. O. Mason, Local Electrical and Dielectric Properties of Nanocrystalline Yttria-Stabilized Zirconia, *J. Mater. Sci.*, **43**, 4684-4692 (2008).

[19] S. L. Gao, X. W. Yang, D. H. Ren, and Q. Z. Shi, Thermochemical Properties of Complexes of Rare Earth Nitrate with Glycine, *Thermochimica Acta*, **287**, 177-182 (1996).

[20] B. D. Cullity, *Elements of X-ray Diffraction*, Addison Wesley Publishing, 2nd edition, 1978.

[21] S. R. Jain, K. C. Adiga, and V. R. Pai Verneker, Flash Combustion Synthesis and Characterisation of Nanosized Proton Conducting Yttria-Doped Barium Cerate, *Combustion and Flame*, **40**, 71-79 (1981).

[22] R. D. Purohit, B. P. Sharma, K. T. Pillai, and A. K. Tyagi, Ultrafine Ceria Powders via Glycine-Nitrate Combustion, *Mater. Res. Bull.*, **36**, 2711-2721 (2001).

[23] W Chen, F. Li, and J. Yu, Combustion Synthesis and Characterization of Nanocrystalline CeO2-Based Powders via Ethylene Glycol-Nitrate Process, *Mater. Lett.*, **60**, 57–62 (2006).

[24] M. Jacquin, Y. Jing, A. Essoumhi, G. Taillades, D. J. Jones and J. Rozière, Flash Combustion Synthesis and Characterisation of Nanosized Proton Conducting Yttria-doped Barium Cerate, *J. New Mater. Electrochem. Syst.*, **10**, 243-248 (2007).

[25] K. Takeuchia, C.-K. Loong, J. W. Richardson Jr., J. Guanb, S.E. Dorris, U. Balachandran, The Crystal Structures and Phase Transitions in Y-doped BaCeO3: Their Dependence on Y Concentration and Hydrogen Doping, *Solid State Ionics*, **138**, 63–77 (2000).

[26] C.-K. Loong, M. Ozawa, K. Takeuchi, K. Ui, N. Koura, Neutron Studies of Rare Earth-Modified Zirconia Catalysts and Yttrium-Doped Barium Cerate Proton-Conducting Ceramic Membranes, *J. Alloys and Compounds*, **408–412**, 1065–1070 (2006).

Fuel Reforming

CARBON DIOXIDE REFORMING OF METHANE ON NICKEL-CERIA-BASED OXIDE CERMET ANODE FOR SOLID OXIDE FUEL CELLS

Mitsunobu Kawano, Hiroyuki Yoshida, Koji Hashino, and Toru Inagaki
The Kansai Electric Power Company, Inc.
11-20 Nakoji 3-chome, Amagasaki
Hyogo 661-0974, Japan

Hideyuki Nagahara, and Hiroshi Ijichi
Kanden Power-tech Co. Ltd.
2-1-1800 Benten 1-chome, Minato-ku,
Osaka 552-0007, Japan

ABSTRACT

Carbon dioxide reforming of methane on nickel (Ni)-ceria-based oxide cermet anode of practical size solid oxide fuel cells (SOFCs) was carried out to clarify the biogas fueled SOFCs. The cell performance and the gaseous composition of the fuel at 750°C were measured under various conditions. Stable cell performance with carbon dioxide reforming was observed. The result of the gaseous composition analysis indicated that methane was effectively reformed by carbon dioxide at the anode. The results show that the anode works effectively for the carbon dioxide reforming, resulting in the stable SOFC performance. It is concluded that carbon dioxide reforming attained sufficient level of conversion for power generation with methane at 750°C on Ni-ceria-based oxide cermet anode of practical size SOFCs.

INTRODUCTION

Solid oxide fuel cells (SOFCs) are expected to be a new promising technology for electrical power generation because of their high electrical efficiency. In general, they must operate at high temperatures near 1000°C. But such high operating temperatures cause many serious problems such as physical and chemical degradations of the SOFCs component materials. Then, intermediate temperature (600-800°C) SOFCs (IT-SOFCs) have been drawing a great deal of attention. The Kansai Electric Power Company Inc. (KEPCO) has been developing IT-SOFCs with nickel (Ni)-ceria-based oxide cermet anode. Ni-ceria-based oxide cermet anode gives high performance due to synergistic effect of enlarging reaction area and increasing paths for ionic and electronic conduction[1].

Another attractive feature of SOFCs is their capacity of internal reforming operation using hydrocarbon fuel because of their higher operating temperatures than the other types of fuel cells[2, 3]. Reforming operation method is generally classified in two ways, which are steam reforming (Reaction 1) and carbon dioxide reforming (Reaction 2). Steam reforming operation has been generally used for a practical SOFC application[4-7]. Carbon dioxide reforming operation, however, has not been adopted as a practical SOFC application. Because steam generator is not necessary and higher electrical efficiency is expected for carbon dioxide reforming operation, it should have more advantage than steam reforming operation. Moreover, methane and carbon dioxide are the main constituents of biogas produced by anaerobic biological waste treatment[8, 9]. Then, investigation of the carbon dioxide reforming operation of SOFCs with methane fuel is considered to be very important for the development of biogas fueled SOFCs.

In this study, carbon dioxide reforming of methane on nickel (Ni)-ceria-based oxide cermet anode of practical size SOFCs was carried out to reveal the electrochemical and the reforming activities for its operation of SOFCs. In addition, external and direct internal reforming operations were compared to clarify its operation in detail. Moreover, carbon dioxide reforming operation was compared with steam reforming operation we have reported previously[10].

$$CH_4 + H_2O \rightarrow CO + 3H_2 \qquad \Delta H^0_{298} = 206 \text{ kJ mol}^{-1} \qquad (1)$$
$$CH_4 + CO_2 \rightarrow 2CO + 2H_2 \qquad \Delta H^0_{298} = 247 \text{ kJ mol}^{-1} \qquad (2)$$

EXPERIMENTAL
Cell fabrication process

Cell fabrication was carried out as the similar method we have reported previously[11, 12]. Optimized $NiO\text{-}(Ce,Sm)O_{2\text{-}\delta}$ (SDC) composite particles prepared by spray pyrolysis were used as raw materials of the anode[11, 12]. NiO-SDC composite particles were reduced into Ni-SDC cermet after the processes of sintering and reducing under a hydrogen atmosphere. $La_{0.8}Sr_{0.2}Ga_{0.8}Mg_{0.15}Co_{0.05}O_{3\text{-}\delta}$ (LSGMC) with 200 μm thickness and $Sm_{0.5}Sr_{0.5}CoO_{3\text{-}\delta}$ (SSC) were selected as electrolyte and cathode, respectively. The diameter of electrodes was 120 mm.

Electrochemical characterizations

As schematically shown in Figure 1, air and fuel were supplied to the center of the cathode and the anode, respectively. This cell stack adopted the seal-less structure in which the unspent residual fuel and air were exhausted at the outer surface of the cell stack. The electrochemical measurement system is schematically drawn in Figure 2. The cell performance was tested at 750°C. Air was supplied to the cathode at a flow rate of 15 ml cm^{-2}min^{-1}. Under the condition of changing operation temperature, hydrogen and nitrogen were supplied to the anode to keep a reducing atmosphere. In power generation, methane and carbon dioxide with various ratios were also supplied to the anode. The flow rate of methane was fixed at 0.75 ml cm^{-2}min^{-1} and the flow rates of carbon dioxide were changed at 3.00, 2.25, and 1.50 ml cm^{-2}min^{-1} to realize methane to carbon dioxide ratio being 1/4, 1/3, and 1/2, respectively. Dry hydrogen was also supplied at a flow rate of 3.00 ml cm^{-2}min^{-1} and the cell performance was tested before and after carbon dioxide reforming of methane. For external carbon dioxide reforming, methane and carbon dioxide were supplied to the pre-reformer containing Ru-Al$_2$O$_3$ reforming catalyst (27 cm^3), prior to introduction to the anode. The temperature of the pre-reformer was set at 750°C. In the case of direct internal carbon dioxide reforming, methane and carbon dioxide were directly supplied to the anode through the by-pass line of the pre-reformer. The composition of the gaseous species in the reformate except for water vapor was measured by micro-gas chromatograph (Varian, SP-4900) before and after the electrochemical reaction. As shown in Figure 3, anode separator had five points for sampling the gaseous species in radial direction, which were 0 mm, 14 mm, 28 mm, 42 mm, and 56 mm, respectively. These values mean the distance from the center hole of the separator. For the electrochemical characterization, I-V characteristics were measured with increasing current density up to recognizing sharp decrease of terminal voltage because of the fuel depletion. AC impedance was measured in the frequency range from 100 kHz to 0.1 Hz with an AC amplitude of 17.7 mA cm^{-2}.

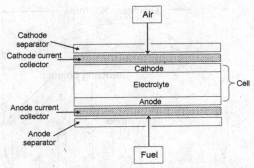

Figure 1. Schematic view of the cross-section of the SOFC cell stack.

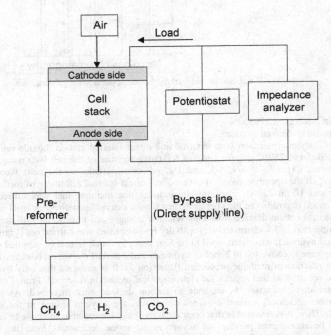

Figure 2. Electrochemical measurement system for SOFC power generation.

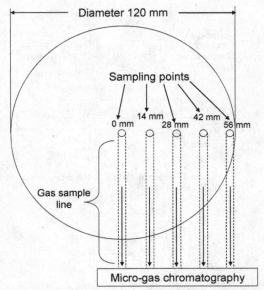

Figure 3. Schematic view of the anode separator for gas analysis.

RESULTS AND DISCUSSION
Electrochemical characterizations

SOFC power generation with external and direct internal carbon dioxide reforming of methane was carried out at 750°C. Figure 4 shows I-V characteristics of the cell with using three kinds of gas compositions; CH_4/CO_2 = 1/4, 1/3, and 1/2. I-V characteristics of the cells were almost the same irrespective of the operation mode of external and direct internal reforming of methane for CH_4/CO_2 = 1/4, 1/3, and 1/2 at 750°C. These results indicate that methane fuel was sufficiently reformed on Ni-SDC anode regardless of the operation runs for the external and direct internal reforming at 750°C. Direct internal carbon dioxide reforming reaction is suggested to proceed sufficiently from Figure 4, then, comparison of I-V characteristics with dry hydrogen fuel was carried out (Figure. 5). The amount of supplied hydrogen was determined to be four times as much as that of supplied methane because 1 mol of methane is converted to 2 mol of hydrogen and 2 mol of carbon monoxide when carbon dioxide reforming reaction of methane proceeded (Reaction 2). It is assumed that only hydrogen and carbon monoxide are used as fuel species for eletrochemical oxidation reactions. From Figure 5, open circuit voltage gradually increased as methane to carbon dioxide ratio increased. As methane to carbon dioxide ratio increased, carbon dioxide, which was the oxygen source in the supplied fuel, was decreased. Then, it is reasonable that open circuit voltage increased as methane to carbon dioxide ratio increased because partial pressure of oxygen at the anode decreased. Under the conditions of high current densities around 0.3-0.4 A cm^{-2}, however, cell voltages were almost the same irrespective of the supplied fuel at 750°C. These results suggest that polarization loss increased a little as methane to carbon dioxide ratio increased under the high current densities.

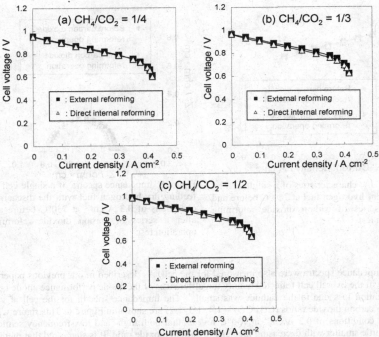

Figure 4. *I-V* characteristics of a single cell at 750°C with external and direct internal carbon dioxide reforming of methane; CH₄/CO₂ = (a)1/4, (b)1/3, and (c)1/2.

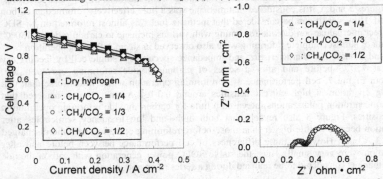

Figure 5. *I-V* characteristics of a single cell by feeding various direct internal carbon dioxide-reformed methane fuels and dry hydrogen fuel at 750°C.

Figure 6. Impedance spectra of a single cell with the discharged conditions of 0.3 A cm⁻² at 750°C using direct internal carbon dioxide-reformed methane fuel.

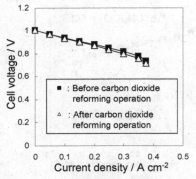

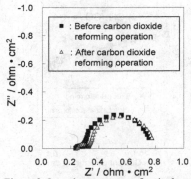

Figure 7. *I-V* characteristics of a single cell by feeding dry hydrogen fuel at 750°C before and after a series of carbon dioxide reforming operation test.

Figure 8. Impedance spectra of a single cell by feeding dry hydrogen fuel with the discharged conditions of 0.3 A cm⁻² at 750°C before and after a series of carbon dioxide reforming operation test.

AC impedance spectra were also measured for the cell. As described in our previous paper[13], the impedance of the overall cell reflected mainly the differences of the anode performance anode because the polarization loss due to the cathode was small[14]. The impedance spectra for the cell at various methane to carbon dioxide ratios (CH_4/CO_2 = 1/4, 1/3, 1/2) are shown in Figure 6. This figure with the discharged conditions of 0.3 A cm⁻² at 750°C reveals that both high- and low-frequency semicircles became a little smaller with decreasing methane to carbon dioxide ratio. It is suggested that methane is effectively reformed by carbon dioxide at lower methane to carbon dioxide ratio. These results from the impedance analysis showed good agreement with those from the data of *I-V* characteristics. Judging from Figures 5 and 6, it is considered that methane was a little effectively reformed as carbon dioxide increased. However, it can be considered that methane fuel was almost reformed on Ni-SDC anode for the direct internal carbon dioxide reforming with various methane to carbon ratios at 750°C. These phenomena of almost complete reforming were also observed in steam reforming operation[10].

Figures 7 and 8 show *I-V* characteristics and impedance spectra of a single cell by feeding dry hydrogen fuel at 750°C before and after a series of carbon dioxide reforming operation test, respectively. From Figure 7, cell performance after reforming operation was a little worse than that before reforming operation at high current densities around 0.3-0.4 A cm⁻². It is considered that activation and concentration polarizations increased a little for carbon dioxide reforming operation at high current densities. Figure 8 also reveals that both high- and low-frequency semicircles after reforming operation became a little bigger than those before reforming operation with the discharged conditions of 0.3 A cm⁻². However, these differences of cell performance between before and after reforming operation were very small. Thus, the stable SOFC power generation with carbon dioxide reforming of methane is considered to be realized during a series of operation test.

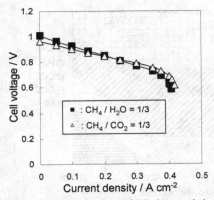

Figure 9. The comparison of *I-V* characteristics for a single cell between direct internal steam-reformed methane (CH$_4$/H$_2$O = 1/3) and carbon dioxide-reformed methane (CH$_4$/CO$_2$ = 1/3) at 750°C.

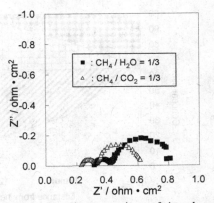

Figure 10. The comparison of impedance spectra for a single cell between direct internal steam-reformed methane (CH$_4$/H$_2$O = 1/3) and carbon dioxide-reformed methane (CH$_4$/CO$_2$ = 1/3) with the discharged conditions of 0.3 A cm^{-2} at 750°C.

The comparison of the cell performance between direct internal steam-reformed methane (CH$_4$/H$_2$O = 1/3) and carbon-dioxide-reformed methane (CH$_4$/CO$_2$ = 1/3) at 750°C was also carried out. The data of *I-V* characteristics and impedance spectra with direct internal steam reforming were previously reported[10]. For carbon dioxide reforming, 1 mol of methane is converted to 2 mol of hydrogen and 2 mol of carbon monoxide when reforming reaction of methane proceeded (Reaction 2). On the other hand, 1 mol of methane is converted to 3 mol of hydrogen and 1 mol of carbon monoxide for steam reforming (Reaction 1). Total fuel species (hydrogen and carbon monoxide) were same for both carbon dioxide and steam reforming, but the ratios of hydrogen and carbon monoxide were different. Their thermodynamically calculated partial pressures of oxygen at the anode and their theoretical terminal voltages were also almost same. Although their theoretical terminal voltages were almost same, open circuit voltage for steam reforming was a little higher than that for carbon dioxide reforming, as shown in Figure 9. This reason was not obvious. However, the voltage for carbon dioxide reforming at high current densities around 0.3-0.4 A cm^{-2} was higher than that for steam reforming. Impedance spectra with the discharged conditions of 0.3 A cm^{-2} shown in Figure 10 also reveals that both high- and low-frequency semicircles for carbon dioxide reforming were smaller than those for steam reforming. These results indicate that carbon dioxide reforming operation provided for the smaller resistance at high current densities. Although it was expected that electrochemical activity for hydrogen was higher than that for carbon monoxide, this assumption conflicted with these results. It is considered that steam might deteriorate Ni-SDC anode activity for steam reforming operation. As a result, carbon dioxide operation is considered to be useful compared with steam reforming operation for practical size Ni-SDC anode.

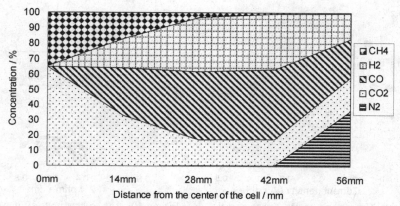

Figure 11. Gas analyzing results in radial direction for a single cell by feeding carbon dioxide-reformed methane ($CH_4 /CO_2 = 1/2$) with the conditions of open circuit voltage at 750°C.

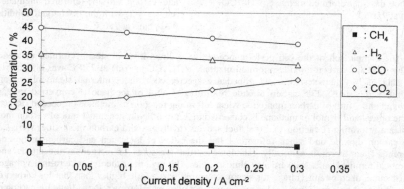

Figure 12. Gas analyzing results at the point of 28 mm for a single cell by feeding carbon dioxide-reformed methane ($CH_4 /CO_2 = 1/2$) with the various discharged conditions at 750°C.

Gas analysis in radial direction of a single cell

As shown in Figure 3, the gaseous composition was measured at five sampling points in radial direction which were 0 mm, 14 mm, 28 mm, 42 mm, and 56 mm, respectively. These values mean the distance from the center of the cell. Figure 11 shows the results of gas analysis in radial direction by feeding carbon dioxide-reformed methane ($CH_4 /CO_2 = 1/2$) with the conditions of open circuit voltage at 750°C. Water vapor could not be analyzed because of the restriction of micro-gas chromatograph equipment. From Figure 11, methane to carbon dioxide ratio at the point of 0 mm was 1/2 as same as supplied ratio, then methane and carbon dioxide were gradually converted to hydrogen and carbon monoxide in proceeding radial direction. The concentrations of each gaseous composition at 28 mm and 42 mm were almost as same as that with thermodynamic calculation. Then, methane was not almost observed at the point of 28 mm and 42 mm. These results reveal that carbon dioxide reforming reaction (Reaction 2) proceeded effectively on Ni-SDC anode under the conditions of $CH_4 /CO_2 = 1/2$ at 750°C. Nitrogen was observed at 56 mm because of the seal-less stack structure. However, oxygen was not almost observed at 56 mm. It is suggested that oxygen combusted hydrogen or carbon monoxide and vanished at the edge of a single cell.

Gas analysis under the discharged conditions was also carried out. Figure 12 shows the gaseous composition at the point of 28 mm for a single cell by feeding carbon dioxide-reformed methane ($CH_4 /CO_2 = 1/2$) with the various discharged conditions (0-0.3 A cm^{-2}). As discussed formerly, methane was almost thermodynamically converted to hydrogen and carbon monoxide with the conditions of open circuit voltage at 28 mm. From Figure 12, it was observed that hydrogen and carbon monoxide decreased as the current density increases. These results suggest that hydrogen and carbon monoxide were effectively used as fuel species for eletrochemical oxidation reactions. In addition, nitrogen and oxygen was not almost observed at the point of 28 mm.

These results of gas analysis reveal that methane was effectively converted to hydrogen and carbon monoxide in radial direction on practical size Ni-SDC anode, and that hydrogen and carbon monoxide were effectively used as fuel species for eletrochemical oxidation reactions.

CONCLUSIONS

Carbon dioxide reforming of methane on Ni-ceria-based oxide cermet anode of practical size SOFCs was carried out. For the comparison of external and direct internal carbon dioxide reforming, I-V characteristics of the cells were almost the same irrespective of the operation mode of external and direct internal reforming of methane. In addition, I-V characteristics of the cells were almost the same irrespective of the supplied fuel for hydrogen and carbon dioxide-reformed methane. It is suggested that methane fuel was sufficiently reformed on Ni-SDC anode for the direct internal carbon dioxide reforming at 750°C. The comparison between steam reforming and carbon dioxide reforming also reveals that the cell performance with carbon dioxide reforming at high current densities was higher than that for steam reforming. Gas analyzing results reveal that methane was effectively converted to hydrogen and carbon monoxide in radial direction on practical size Ni-SDC anode, and that hydrogen and carbon monoxide were effectively used as fuel species for eletrochemical oxidation reactions. It is concluded that carbon dioxide reforming attained sufficient level of conversion for power generation with methane at 750°C on Ni-ceria-based oxide cermet anode of practical size SOFCs.

REFERENCES
[1]A. Atkinson, S. Barnett, R. J. Gorte, J. T. S. Irvine, A. J. Mcevoy, M. Mogensen, S. C. Singhal, and J. Vohs, Advanced anodes for high-temperature fuel cells, *Nature Materials*, **3**, 17-27 (2004).
[2]A.L. Dicks, Advances in catalysts for internal reforming in high temperature fuel cells, *J. Power sources*, **71**, 111-122 (1998).
[3]E. P. Murray, T. Asai, and S. A. Barnett, A direct-methane fuel cell with a ceria-based anode, *Nature*, **400**, 649-51 (1999).
[4]H. Kishimoto, T. Horita, K. Yamaji, Y. Xiong, N. Sakai, M. E. Brito, and H. Yokokawa, Feasibility of n-Dodecane Fuel for Solid Oxide Fuel Cell with Ni-ScSZ Anode, *J. Electrochem. Soc.*, **152**(3), A532-8 (2005).
[5]H. Kishimoto, K. Yamaji, T. Horita, Y. Xiong, N. Sakai, M. E. Brito, and H. Yokokawa, Reaction Process in the Ni-ScSZ Anode for Hydrocarbon Fueled SOFCs, *J. Electrochem. Soc.*, **153**(6), A982-8 (2006).

[6]I. Gavrielatos and S. Neophytides, High Tolerant to Carbon Deposition Ni-based Electrodes under Internal Steam Reforming Conditions, *ECS Trans.*, 7(1), 1483-90 (2007).

[7]N. Laosiripojana, S. Assabumrungrat, and S. Charojrochkul, Steam Reforming of Ethanol over Ni on High Surface Area Ceria Support: Influence of Redox Properties on Catalyst Stability and Product Selectivities, *ECS Trans.*, 7(1), 1717 (2007).

[8]J. V. herle, Y. Membrez, and O. Bucheli, Biogas as a fuel source for SOFC co-generators, *J. Power sources*, 127, 300-12 (2004).

[9]J. Huang, and R. J. Crookes, Assessment of simulated biogas as a fuel for the spark ignition engine, *Fuel*, 77, 1793-1801 (1998).

[10]M. Kawano, M. Matsui, R. Kikuchi, H. Yoshida, T. Inagaki, and K. Eguchi, Steam reforming on Ni-samaria-doped ceria cermet anode for practical size solid oxide fuel cell at intermediate temperatures, *J. Power Sources*, 182, 496-502 (2008).

[11]M. Kawano, K. Hashino, H. Yoshida, H. Ijichi, S. Takahashi, S. Suda, and T. Inagaki, Synthesis and characterizations of composite particles for solid oxide fuel cell anodes by spray pyrolysis and intermediate temperature cell performance, *J. Power Sources*, 152, 196-9 (2005).

[12]M. Kawano, H. Yoshida, K. Hashino, H. Ijichi, S. Suda, K. Kawahara, and T. Inagaki, Studies on synthetic conditions of spray pyrolysis by acids addition for development of highly active Ni-SDC cermet anode, *Solid State Ionics*, 177, 3315-21 (2006).

[13]M. Kawano, T. Matsui, R. Kikuchi, H. Yoshida, T. Inagaki, and K. Eguchi, Direct Internal Steam Reforming at SOFC Anodes Composed of NiO-SDC Composite Particles, *J. Electrochem. Soc.*, 154(5), B460-65 (2007).

[14]T. Ishihara, M. Honda, T. Shibayama, H. Minami, H. Nishiguchi, and Y. Takita, Intermediate Temperature Solid Oxide Fuel Cells Using a New LaGaO$_3$ Based Oxide Ion Conductor -I. Doped SmCoO$_3$ as a New Cathode Material-, *J. Electrochem. Soc.*, 145, 3177-83 (1998).

Author Index

Author Index